Savannas: A Very Short Introduction

VERY SHORT INTRODUCTIONS are for anyone wanting a stimulating and accessible way into a new subject. They are written by experts, and have been translated into more than 40 different languages.

The Series began in 1995, and now covers a wide variety of topics in every discipline. The VSI library now contains over 450 volumes—a Very Short Introduction to everything from Psychology and Philosophy of Science to American History and Relativity—and continues to grow in every subject area.

Very Short Introductions available now:

Available soon:

For more information visit our website

www.oup.com/vsi/

Peter A. Furley

SAVANNAS

A Very Short Introduction

OXFORD
UNIVERSITY PRESS

Great Clarendon Street, Oxford, OX2 6DP,
United Kingdom

Oxford University Press is a department of the University of Oxford.
It furthers the University's objective of excellence in research, scholarship,
and education by publishing worldwide. Oxford is a registered trade mark of
Oxford University Press in the UK and in certain other countries

Published in the United States of America by Oxford University Press
198 Madison Avenue, New York, NY 10016, United States of America

British Library Cataloguing in Publication Data
Data available

Library of Congress Control Number: 2016933313

ISBN 978-0-19-871722-5

Printed and bound by
CPI Group (UK) Ltd, Croydon, CR0 4YY

Contents

Acknowledgements

My grateful thanks to James Ratter of the Royal Botanic Garden, Edinburgh (RBGE), for his collaboration over more years than we both care to remember, and for his careful reading of the manuscript. Thanks also to colleagues both at the RBGE and at the University of Edinburgh, and to many generations of students working in the UK and in the field overseas for valuable help and insights into the issues facing different savanna countries. Particular thanks to Sarah O'Donogue for help with illustrations, and Dimitrios Michelakis and Neil Stuart for guidance on current remote sensing technologies.

In this account, the definition of savanna is not taken to include the temperate grasslands of higher latitudes, or the grassy systems that occur at altitudes over around 1,000 metres above sea level. The spelling *savanna* is taken from the Spanish and was originally derived from the native Carib peoples of Middle America.

List of illustrations

Savannas

List of tables

Chapter 1
Savanna landscapes

What are savannas?

The term 'savanna' frequently conjures up images of vast grassy plains interspersed with umbrella-topped trees and populated by herds of large grazing animals. These images, characteristic of Africa, offer only a partial insight into the extraordinarily rich and varied landscapes that make up world savannas. Such landscapes are amongst the most dynamic and diverse *biomes* (or biological life zones) that cover the Earth's surface. They represent a fascinating, colourful, and abundant resource that has performed a key role in the evolution of human beings and continues to tax our ingenuity in balancing resource use against conservation in a world of rapidly growing population.

Savannas are the most widespread form of vegetation in the tropics and sub-tropics, making up around half of the surface area. They cover well over 20 per cent of Australia depending upon how they are defined, around 45 per cent of South America, and more than 50 per cent of Africa. There are patches across India, South-east Asia, and the Pacific islands, and savannas extend up through Central America and the Caribbean towards North America. They are most simply described as areas covered by a nearly continuous grassy layer, interspersed with trees and shrubs of varying densities and heights. The nature of savannas will be explored later, but attempts to give a precise definition are fraught with difficulty because savanna vegetation is frequently a mosaic of trees, grasses, and shrubs. The proportions of these not only change rapidly over time but also include patches that are not strictly categorized as savanna. Consequently generalizations that seem appropriate at a global scale are not necessarily apposite at a regional or local level.

The notion of a savanna landscape is a conceptual image as well as a visual panorama. Despite widespread appreciation of what constitutes such a landscape, there have been several different approaches to interpreting its meaning. Most people have an idea of a mapable unit where land, people, and history combine to give a unique image and memory of place. Landscape comprehends scenery, wildlife, and human activities. It includes vegetation that is unarguably recognized as savanna, as well as tracts of different formations and their constituent biological organisms. Issues of scale and the level of technicality also impinge upon definition but, whichever interpretation is appropriate, it has been estimated that around one-fifth of the world's population lives in or near savanna regions. Because savannas are typically located between the consistently wet and consistently dry regions of the tropics, they inevitably overlap, at a broad scale, with the peripheries of forest on the one hand and arid regions on the other. They merge with temperate grasslands and woodlands, and may also abut aquatic *ecosystems* (smaller-scale ecological units) associated with river valleys, lake margins, wetlands, and coastal areas. Savannas are also often criss-crossed by rivers, giving rise to gallery forests and shallow damp depressions forming seasonally flooded and more marsh-like environments such as the Pantanal in South America or the Okavango Delta in southern Africa. They may be interrupted by influential geological features, such as volcanic deposits or limestone outcrops that shape the nature of soils supporting different types of vegetation. In this way savanna landscapes typically include tracts of differing habitats, which nonetheless make up the total multifaceted picture that characterizes a rich and intriguing biome.

Where are the savannas?

The global distribution of savannas (see Figure 1) illustrates the broad pattern to either side of the lines of the tropics. However, such a large-scale view conceals both the diversity and the far-reaching character of the plant and animal life. Most savannas reflect strong seasonal climates but their boundaries have shifted over time with climatic changes and, more dramatically, since the onset of human disturbance. Many savannas are today better termed *derived savannas*, where the vegetation resembles true savanna superficially but may relatively quickly revert to some other form given respite from disturbance. In this way we may refer to the true savanna as *climax vegetation* (in other words, the type of vegetation that may theoretically occur given equilibrium with climate and other determining factors over a long period of time). This attractive supposition is hardly ever achieved, since savannas are constantly changing and there is debate as to whether savannas are simply transitional landscapes perpetually in the process of transformation.

The world picture does nevertheless show several characteristic features. Savannas are clearly widespread in tropical and sub-tropical latitudes and comparisons with maps of world climate demonstrate that they occur over areas with highly seasonal and often irregular rainfall. Savannas are found predominantly in the southern hemisphere, with the major realms located in Africa, large stretches of South America, and Australia. Africa holds the largest areas of savanna in the northern hemisphere but savanna landscapes extend into Central America and the Caribbean with scattered tracts throughout India, South-east Asia, and numerous other less extensive localities in the sub-tropics. At a local level, differing plant associations make up a mosaic of vegetation formations whose features can vary over a few metres. Each of these scales has its own distinctive character and contributes to the overall diversity of the savanna landscape.

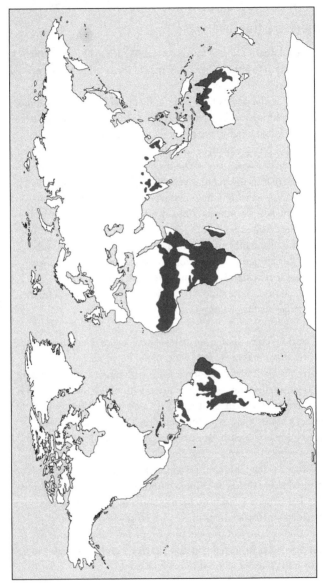

1. The world distribution of savannas.

Although savannas result mainly from similar temperature and rainfall conditions throughout the world, they show distinct differences in fauna and flora. The similarities between savanna and neighbouring biomes within each of the continental ranges can be greater than the similarities between different world savanna regions. For example, the vegetation of the Brazilian savannas (or *cerrados*) shows greater affinity with the flora of the Amazonian forest than to the savannas of Africa or Australia.

Marked seasonal variation is characterized by unpredictably long wet or dry spells. The dryness tends to restrict tree growth and as a result there is a range of vegetation that stretches from sparse grassy and shrubby formations near the drier margins to nearly continuous tree canopies at the wetter transitions (see Figure 2). Furthermore although savannas are typically found over the continental plateaus of the southern hemisphere, they are also found over varied tropical environments from sea level to altitudes over 1,000 metres (m). In addition, they are predominantly underlain by coarse-textured, weathered, and infertile soils that have accounted for their lack of exploitation for cultivation in the past. At greater altitudes and at latitudes further north and south of the tropics, they may merge into more typically temperate grasslands and woodlands, which, however, lack the same characteristic grass species.

Savanna landscapes

Landscape captures the total picture that can be observed at any one place at a given period of time. This is usually underpinned by a combination of topography and vegetation. The structure and general composition of the plant cover is often the feature that catches the eye, and is at first sight similar in appearance and function throughout the world's savannas. However, a closer look reveals changes in *physiognomy* or appearance, and *phytosociology* or plant associations. These result in part from different evolutionary pathways determined by factors such as continental drift and the present-day positioning of the savanna zones within seasonal

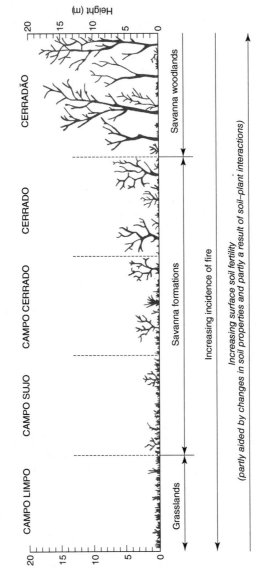

2. Characteristic forms of savanna vegetation and the typical composition of the Brazilian *cerrados*.

climates. More spectacular differences are seen in the variety and numbers of animals, and in the movement and settlement of humans.

Landscape reflects the sweep of topography, influenced by the underlying rock composition and structure. In savannas the topography can change from rolling or steep-sided hills to smooth undulating plateaus and level plains. The steepness of the relief also controls the flow of water over the surface and the drainage through the soil, and in turn these processes shape the form and variety of plants. The vegetation is made up of perennial plants with permanent above-ground structures; perennials with seasonal above-ground structures but often with underground perennating structures; and annuals, with a marked lack of ephemeral species (see Chapter 3).

The frequency and intensity of fire complicates a smooth succession of plant colonization and all savannas have adjusted to burning at regular, frequent, or periodic fire intervals. They have been termed 'fire-adapted landscapes' because of the constant pattern of burning and recovery. With frequent burning and long dry seasons, tree and shrub colonization is inhibited and grasslands become dominant (see Figure 3(a)). Where there is greater rainfall and shorter periods of dryness, more woody species can colonize and persist. Eventually a tree cover may become so well established that it can resist burning and resembles woodland. It is only defined as 'savanna woodland' by the persistence of a grass and herbaceous ground cover that becomes more evident in the drier parts of the year when the taller vegetation loses its leaves, as in the *miombo* of southern Africa, the *cerradão* of Brazil, or some of the eucalypt-dominated savannas in Australia (see Figure 3(b)).

As a consequence of fire, savanna plants and dependent animal and micro-organic activities are in a continuous process of alteration, and this is further accentuated by the relative unpredictability of climate and the impact of human disturbance. The result of these continual pressures is often a patchwork of different plant associations.

3(a) Savanna grassland landscapes at Amboseli, Kenya, with mountains of the Rift Valley in the background; (b) Woody savannas in the Brazilian *cerrado*, showing the nature of the herbaceous ground cover and arboreal termite nests.

Geological evolution and climate change

Savannas are mostly found in the southern continents where there are the largest land masses. These land masses are remnants of the ancient supercontinent of Gondwana and reflect its geology with interconnections past and present that link the current lands together (see Figure 4). The nature and distribution of savanna plants and animals is closely allied with the evolution of the continents, and thus with the composition and structure of the underlying rocks together with the characteristically poor soils.

Gondwana ('forest of the Gonds'), was named after the area in India where the geological chronology was first revealed in the 19th century. It contained a sequence of sediments dating from the Permian to Cretaceous Periods c.300–65 Mya (million years ago) that were later found to be similar, in composition and fossil remains, to sediments found in the other southern continents (see Table 1). Over

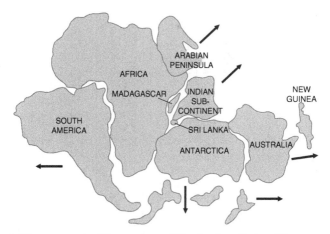

4. The movements of the continents following the splitting of the ancient land mass of Gondwana.

Table 1 The geological timescale

Era	Period	Epoch	Age (Mya)	Significant events
Cenozoic	Quaternary	Holocene	0.01–present day	The Anthropocene; end of the Ice Age; and dominance of humans
65.5 Mya–present day		Pleistocene	1.6–0.01	Ice ages and earliest *hominins*
	Tertiary	Pliocene	5.3–1.6	Early humanoids in Africa
		Miocene	23.7–5.3	Emergence of grasslands
		Oligocene	36.6–23.7	
		Eocene	57.8–36.6	
		Palaeocene	68.5–57.8	Earliest large mammals
Mesozoic	Secondary	Cretaceous	144–65.5	Earliest flowering plants
245–65.5 Mya		Jurassic	208–144	Old sub-continent (Pangaea) divides into Laurasia and Gondwana
		Triassic	245–208	
Palaeozoic	Primary			
570–245 Mya				
Pre-Cambrian				
c.4,000–570 Mya				

this long period of time, the southern continents are believed to have contained broadly analogous floras although there were regional differences. The break-up of Gondwana started some 180 Mya in the Jurassic and continues today with fragments moving slowly northwards. It seems that the first rifting of the supercontinent happened in the west, notably South America splitting from Africa, and extended eastwards during the Cretaceous. The result of all these movements over time was the gradual shifting of varied rock formations into what are now warmer latitudes. It is the similarity in seasonal climate over recent time that strongly influences the distribution and pattern of present day vegetation. Before and after the rifting, the vegetation was dominated by conifers and ferns, but *angiosperms* or flowering plants were established before the final break-up and diversified in the Cretaceous Period. In this way it can be seen that latitudinal controls have had a marked effect on floras through time. While the floras are similar in appearance and function, and plant life was able to spread and colonize with worldwide distributions, animal life was less able to cross water or other physical and chemical barriers; the different fragments became isolated from one another and so evolved characteristically different forms of wildlife.

Despite the importance of geology, the distribution of savannas is usually distinguished on the basis of their climatic characteristics, with marked wet and dry seasons. However, there are several typical savanna landscapes that lie outside what might normally be considered as a characteristic savanna climate. This occurs, for example, where local factors override the influence of climate, as in the patches of savanna found within the Amazon rainforest, where exceptionally sandy soils have inhibited tree growth. Equally, there are tracts within the savanna, experiencing the same overall climatic determinants, that do not contain savanna vegetation. Examples include dry deciduous forest located within the Brazilian savanna that result from more fertile calcareous soils derived from exposures of underlying limestone.

The key features of the evolutionary story that occurred in savannas are the rise of primates from around fifty-six Mya, the appearance and expansion of grasses from early Eocene some thirty-five Mya, and the evolution of ancestral humanoids in Miocene about twenty Mya. In terms of geological time, the savannas are relatively recent arrivals, with the expansion of grass-dominated landscapes and the differentiation of vertebrate communities developing from the Tertiary Period onwards.

Biological richness

There are marked differences in biological richness throughout savanna regions. The landscapes may vary from open grasslands filled with large grazing animals and iconic predators (forming a distinctive food chain) as in many parts of Africa; to open grasslands with hardly any large animals (stretches of South America); or open savannas with fewer specialized animals (Australia). It is well known that species richness varies with area, and consequently comparisons between different continental realms or tracts with a given region can only be meaningful if the areas are of similar size. Nevertheless there are broad comparisons that are useful at this stage. The South American savannas are, for example, generally more moist, with a greater proportion of evergreen plants, a large rodent population, and few large animals; the African savannas are characteristically grasslands with a pattern of acacia trees and numerous large animals; and Australia has typically more open eucalypt woodlands and grasslands with distinctive marsupial animal populations. Overall the fauna and flora of the major continental savannas share remarkably few life forms at species level.

'Biodiversity' is the collective term for the total numbers and variety of life. While the richest biomes in terms of species can be found in some rainforests and coral reefs, there are a number of savanna regions that can be termed *biological hotspots*. These are concentrations of organisms that have been recognized as important for their outstanding natural richness and with a pressing need for conservation. There is justifiable concern that this biodiversity is being threatened, partly from natural causes but also from increased human disturbance. For instance, the Brazilian *cerrados*, which were practically undeveloped before the construction of the new capital, Brasília, in the late 1950s and early 1960s, have been converted into land for cultivation and pasture at an extraordinarily rapid pace. Over 1,000 species of trees and shrubs and over 12,000 vascular plants (those with tissues for conducting water and minerals and the products of photosynthesis) have been recorded in the Brazilian *cerrados*. Levels such as these rival some of the moist evergreen forests. Similarly in West Africa the total plant species count has been estimated to be only a little short of that of neighbouring rainforest. Animal variety is perhaps better known and African savannas are remarkable for their profusion of wildlife. In much of Africa, humans and wildlife have co-existed for thousands of years and, until relatively recently, a reasonable balance has been struck; only over recent decades have major discordances been experienced that threaten the futures of savannas.

Savannas and human evolution

There have been numerous theories on the role of savannas and open landscapes in shaping the emergence and spread of human populations. One line of argument has been that it was the migration of primates from forest areas to savanna woodland and eventually to more open savannas that influenced the physical adaptation and social organization separating early from later humans (*hominins*). The early humans were hunter-gatherers who ranged over a much larger territory than the primates. The greatest variety of primates lived in Africa and this continent is generally considered to be the origin of the human species. Fossil apes living around ten to nine Mya display a measure of ground adaptation and some groups moved out into more open landscapes, for example acacia woodlands. Humanoids evolved in the late Miocene, seven to five Mya and *hominins* emerged in the late Tertiary to Quaternary. *Homo erectus* (considered the first true human) appeared around 1.8 Mya. At this stage there were possibly only some 10,000 individuals throughout Africa where there were likely to have been several points of development. The closest relations of modern humans probably migrated out of Africa five to two Mya.

A further theory argues that it was not simply a matter of physical adaptation to savanna-like conditions but the evolution of social organization and behavioural patterns that allowed humans to move more freely in non-forested environments. There was clearly an intricate web of processes influencing the physiognomy and social character of the early pioneers venturing out into more open landscapes. An interesting line of evidence suggests that the diet of early humans had a grass component, although they lived relatively close to water. The first use of fire has been postulated for a site in Kenya at around 1.6 Mya using evidence from burnt soils. Climate change may also have been a contributing factor with some evidence for a contraction in the forest cover in the late Miocene and Pliocene commensurate with an expansion of grasslands. This would have provided new challenges and opportunities, as suggested by the greater use of tools.

Developments in DNA (deoxyribonucleic acid) sequencing have added a fascinating insight into these evolutionary processes. The genetic record of life is preserved in the *genomes* (an organism's complete set of DNA including all of its genes) of living species, which are passed down over generations. This permits tracking of genetic characteristics and can generate phylogenetic trees, or chronological diagrams of evolution. Our understanding of the evolution and migration of humans is likely to be greatly increased in coming years as the sampling intensity increases and more evidence becomes available. The emergence of humans and their gradual transformation of savanna landscapes may have been the outcome of a change in climate and expansion of savanna-like conditions, along with anatomical changes from four to two legs, which freed the use of hands, together with a use of fire and a change of diet, including plant and animal sources (see Chapter 5).

The sequence of this book deals first with the way in which the biome has been shaped by physical features and by the dynamic changes produced by determining environmental and biological processes. The narrative then moves on to a consideration of the differences between the world's different savanna regions with their varied plant and wildlife and different histories of human occupation. Savannas are being consumed at an unprecedented rate, and the issues of development and conservation are more urgent than ever before. The final chapters examine the ways that people have been shaped by savannas, and have currently settled or utilized their environment, with a consideration of current issues and future directions.

Chapter 2
Shaping the savannas

Although savanna landscapes appear to be stable, in reality they are restless panoramas in a state of continuous change. Some of the determining factors shaping the view at any given locality are relatively short-lived. Fires, storms, seasonal variations, and some wildlife disturbances, such as elephant damage or grazing pressures, and many human activities last for periods of days, weeks, or a few months. Other factors influence the landscape over much longer stretches of time, such as the development of the underlying soil profile and more gradual processes of climatic change.

Climatic drivers

At one time savannas were thought to be so constrained by climate that the term 'savanna climate' was coined. The notion of a type of climate specific to savanna landscapes has since been discredited, as there are many tracts of savanna that do not conform to the characteristic parameters of a savanna climate. However, most savanna landscapes are essentially the product of alternating wet and dry seasons in regions having constant warmth with mean temperatures rarely lower than 18°C at sea level. There are only small fluctuations in solar radiation, and daily temperatures typically range between 25 and 30°C. The dry seasons may persist from between around three to over seven months in any year. At higher altitudes there are greater daily fluctuations and it is not uncommon to find frosts. These can have a marked damaging effect on vegetation (e.g. in the savanna woodlands of the Highveld at 1,200 m in central Zimbabwe), and can even result in light snow (causing considerable discomfort to human residents unaccustomed to such temperatures, as in some of the satellite towns around Brasília at over 1,000 m).

Vegetation responds mainly to differences in annual rainfall, particularly its spatial and temporal distribution and consistency over a period of years. Rainfall over the world's savanna regions ranges from less than 300 mm to over 2,000 mm per year, with high humidity in wetter periods. The evaporation rates in drier times of the year often exceed total moisture availability. The amount of rain directly influences the type of vegetation, and a study of several hundred sites across Africa has shown that the proportion of woody vegetation increases linearly with rainfall up to around 650 mm per year. The impact of these variations on vegetation also depends upon the intensity, reliability, and drainage as much as the total amount of rainfall. This can be summarized by describing the plant available moisture, which is reflected in both the precipitation and water balance to plant rooting depth, and in simple climatic diagrams (see Figure 5).

The seasonal patterns reflect the broad climate picture of the tropical zone. Because of the Earth's rotation in relation to the Sun, the overhead Sun at the Equator means that the same amount of radiant heat is concentrated over a smaller area than that received at higher latitudes and temperatures are generally higher. With low pressure around the Equator and high pressures over the lines of the tropics, the resulting wind patterns push air masses from high to low areas. Combined with the Coriolis force resulting from the spinning of the Earth's axis, this generates the well known trade winds between the lines of the tropics. A returning higher level air flow lifts warm, humid air from the Equatorial regions and returns a hot drier downflow over the tropics.

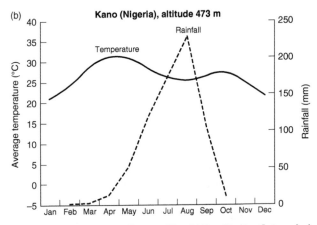

5. Typical savanna climates illustrated by: (a) Brasília (South America);
(b) Kano (West Africa);

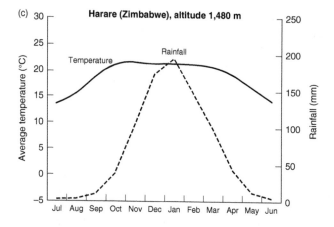

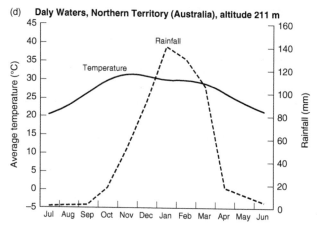

5(c) **Harare (southern Africa); and (d) Daly Waters (northern Australia).**

As well as these global processes, there are changes caused by the annual movements of the intertropical convergence zone (ITCZ) and the impact of El Niño events. The ITCZ is a low pressure region looping the Earth close to the Equator where the trade winds from the northern and southern hemispheres meet. The convergence zone in July tends to be further north, swinging from Central America, through northern South America across Africa and the Arabian Peninsula, North India, and China. In January it is located across South America from Colombia to the north of Brazil, across the Atlantic to West Africa, and then swinging south across central Africa touching the north of Madagascar then east to Indonesia and the northern parts of Australia. As the warm, moist air in this convergence zone rises, it expands and cools triggering thunderstorms and bringing heavy rain. The annual shifts in the ITCZ accentuate the wet and dry seasons, and can result in flooding or droughts.

The El Niño events (El Niño and Southern Oscillation, or ENSO) can augment this impact. They are very large-scale climatic disturbances which originate in the tropical Pacific and occur approximately every three to seven years and can persist for a few months or a few years. El Niño is the warm phase of ENSO events associated with a belt of warm ocean water. The cool phase, known as La Niña, experiences surface sea temperatures cooler than normal and brings about marked differences in pressure between the east and west Pacific. The ENSO events are of such magnitude that they cause global changes in temperature and rainfall, as well as accentuating the variability of climate in savanna regions. Although El Niño conditions vary from one episode to another, there are a number of relatively consistent patterns. For example, the northern part of South America tends to be drier than average, along with most of the Australian and East Asian savannas, most of the Indian sub-continent and a latitudinal band across Africa from Gambia to the Horn of Africa, and a large part of southern Africa. By contrast, a small area to the south of Brazil, the southern tip of India plus Sri Lanka, and a considerable area of the East African savannas are distinctly wetter than normal. Such large-scale climatic changes, along with associated smaller seasonal variations, have a significant influence on natural vegetation and severe impacts on agricultural production and the economies of developing countries. The implications of these ENSO events are being monitored by a number of global institutions such as the CCAFS (a collaborative organization dealing with Climate Change, Agriculture and Food Security. See the websites in the Further Reading section of this volume).

Severe storms and strong winds also have dramatic effects on the vegetation. The destructive actions of cyclones and heavy monsoon rains affecting areas around the Indian Ocean and northern Australia, or typhoons in the Pacific region, or hurricanes in Middle America, not only damage the vegetation but can also alter the appearance of the entire landscape. In Belize, for example, the effect of a series of hurricanes over approximately thirty-year intervals has been to change the visual landscape in the Maya Mountains from fairly dense pine savanna woodland to open grassland with a few remnant trees. Yet with time, plant succession tends to re-establish the vegetation structure best adapted to the climate and environmental pressures—until the next episode of disturbance. In the case of the Maya Mountains this was illustrated by the successive stages of re-growth from grassland to pine-oak savanna woodland. Whilst catastrophic storms are destructive, they may also have a positive side in that they open up the tree canopy allowing in light and encouraging the colonization of a range of other plants thereby increasing diversity.

At a more local level, the nature of the vegetation cover also influences the water balance. Where there is a well covered ground surface, with a dense mixture of trees, shrubs, and herbaceous plants, the incoming rainfall is held up and partly absorbed by the vegetation and soil organic matter. This buffering effect releases water slowly to the soil and plant roots and protects the ground surface from raindrop impact and potential erosion. The plant cover also influences temperature and humidity, and the ground surface of savanna woodlands tends to be cooler and moister than adjacent open grassland. Where the vegetation cover is disturbed, either by natural events or by human clearance, the ground surface rapidly dehydrates and organic matter decomposes in dry periods, while the soil becomes exposed to rainfall and leaching in wet spells. In both situations, the topsoil becomes vulnerable to weathering and soil fertility suffers.

The role of vegetation

The pattern of woodlands, bush, and thicket or open grassland provides the strongest visual impression of the savanna landscape. The proportions of these different life forms vary across many savanna regions even though the predominant images are of widespread grassy plains, such as those seen in eastern Africa or in the Llanos of Venezuela and Columbia, or over the great expanse of shrub-like semi-arid savanna in Australia. Each life form has its own set of mechanisms for adaptation. In most savannas the major life forms may be simplified into trees and shrubs; epiphytes or air plants which cling to the trees; ground-surface plants; plants that die back in the dry season but regenerate from underground tubers in the wet; and plants that cope with marshy or seasonally flooded habitats.

Each plant life form has developed its own set of mechanisms for adaptation. Although trees are infrequent in some of the grassland savannas, and even where they are present in low numbers, they represent conspicuous features of the landscape. Trees rarely exceed 15 to 20 m in height even in savanna woodlands where, from a distance, the canopies appear to be touching and more closely resemble forest. The closed canopy appearance is most obvious in the wet season when trees are in full leaf, and the true character of the savanna is only apparent in the dry periods. Even in the densest savanna woodland, light can penetrate to the ground surface during the dry periods allowing grasses and other herbaceous plants to flourish. In more open conditions, the trees often develop umbrella-shaped crowns to gain the maximum benefit from solar energy and, in progressively drier environments, may take on a stunted appearance. The structure of trees varies from the flat-topped acacias typical of much of Africa (see Figure 6(a)) to the gnarled and twisted forms developed as a result of fires with adventitious resprouting of branches in the denser forms of *cerrado* (see Figure 6(b)). Tree densities also vary with mean annual rainfall from sporadic trees and bushes at the arid margins to almost continuous tree cover in savanna woodlands such as the *miombo* in southern Africa or the *cerradão* in Brazil.

Bush and shrub savannas also possess unusual adaptations. Many of the bushes and low trees such as the acacias are armoured with sharp thorns to deter grazing and some have symbiotic associations with organisms that dissuade attack on leaves. Many of the shrubs also resist fires by producing root tubers that survive the burning of the upper parts of the plant and act as storage organs for water and nutrients. Grasses and other plants typical of the herbaceous component of savannas have also developed adaptations to the environment. They have to survive in a regime of extreme heat and dryness with a high risk of burning. Annuals by definition die down during the dry season but set large quantities of seed that are dispersed widely in winds. Biennials pass through their growing cycle over two years, while perennials resist the dry times of the year by dying down at the surface and recovering when moisture becomes available. The tough protective leaf sheaths at the base of the plant and the extensive rooting systems (calculated at many kilometres in total for some plants) enable a rapid regeneration. Some plants are restricted in height by dryness or fire, but would grow taller given more favourable conditions. Disturbances such as fire and human activity are overcome through time, but as with frequent burning,

6(a) Flat topped acacias (umbrella trees) in the Serengeti National Park, Tanzania; (b) Contorted structures of trees in the Brazilian *cerrado*.

continuous human disturbance weakens the ability of plants to regenerate.

Although it is readily apparent that vegetation structure is constantly varied over the landscape, it is often thought that plant composition is similar over large areas. Research over the last few decades has revealed that whereas some savannas have a species richness that rivals evergreen forest, others have a far less varied flora.

Forest–savanna boundaries

Boundaries between different vegetation formations and plant communities are always of interest because they act as testimony to the balance of processes pitting one ecosystem against another. In the case of the forest to savanna boundary, the frequently sharp divide between the two (sometimes over as little as a few metres), often indicates that previous burning has lapped up to the forest edge and destroyed the vulnerable vegetation. Conversely where burning has been absent for a few years and there is sufficient moisture, the wooded fringe of forest advances over the savanna. The tendency for wooded vegetation to prevail is counteracted by the incidence of fire. At the present time there is evidence in many parts of the world that the forest frontier is competing strongly and overriding areas that were previously more open. This is most likely to be the result of gradual climatic change, where the equilibrium has moved to a moister phase and forest can compete. This is very evident in Brazil, where the southern fringes of forest have advanced in living memory for tens of kilometres in places. A further striking example can be found in the scattered patches of savanna within the Amazonian forest, where characteristic savanna indicator trees such as the sandpaper tree (*Curatella americana*, a genus with one unique species) and the genus *Byrsonima* (with well over a hundred species of trees and shrubs) are being progressively engulfed by the neighbouring forest (see Figure 7). Similar advances of forest into previously more arid

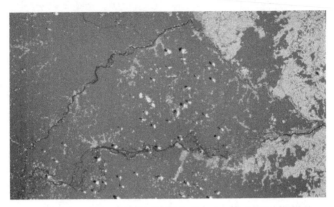

7. Satellite image (LANDSAT TM) of the forest–savanna boundary in northern Amazonia. The pattern of vegetation on Maracá Island in Roraima, Brazil, is shown, with the lighter coloured savanna areas to the east (right) and residual savanna patches within the forest.

regions have been reported for a number of places in Africa (such as southern Angola). Over sloping topography, where savanna vegetation may dominate on the higher ground and forest on the lower damper sites, burning often shapes the entire boundary. Some plants, for example the sedge in Central America, known as the shaving brush (genus *Bulbostylis*) on account of its shape, act as incendiaries, catching fire in the upper savanna and rolling down the slope to ignite the vegetation below.

Fire and the savanna landscape

Fires have been recorded from charcoal remains dating back to the rise of grasslands in the Pliocene. Since that time, savanna plant structure and composition has been shaped by and adapted to burning. Wildfires resulting from lightning strikes, or even through the igniting action of sunlight on very dry combustible organic matter, have been dated to periods long before the arrival of humans or primates capable of manipulating fire. In recent time, the majority of fires are probably manmade, although

storm-generated fires are still an important factor. The role of
humans in starting fires has a very long history, using fire as a
tool for flushing out game, for clearing land, reducing pests and
diseases, and for social organization such as access, settlement,
cooking, and the beginnings of pastoralism and cultivation.
Although the part played by the indigenous people in Australia is
a good and perhaps contentious example of fire changing the
original landscape, there is an established link from pollen and
charcoal evidence between the arrival of the earliest humans
in Australia over 40,000 years ago and the expansion of
fire-adapted vegetation.

Savanna vegetation shows remarkable adaptability to the
regular or periodic occurrence of burning. Grasses afflicted by
frequent fires sweeping across the landscape are burnt off
above ground level. However, they are protected by relatively
inflammable sheaths at the base and by the very extensive root
systems which can delve down into more favourable moist
layers below the surface. Some of these grass root systems are
so prodigious that they have a biomass far greater than the
biomass of the vegetation above ground. Other life forms have
equally inventive measures to protect against burning. Many of
the shrubs, for example, may be badly burnt at the surface but
possess large lignotubers (*geoxyles*) that enable the plant to
bounce back into life at the end of the fire (see Figure 8(a) and
(b)). Even the trees have developed adaptations and many have
the ability to resprout from root shoots, and the tuberous
structures can be so extensive that they have been referred to as
'underground forests'. Since most fires sweep rapidly across the
landscape to a height of a few metres, tree canopies are better
protected once the plants have reached a safer height. Other
trees have developed bark protections that resist burning. For
example, in the Brazilian *cerrado* many of the trees have thick and
cork-like protective barks and are frequently strikingly fissured,
which is believed to assist in the rapid dissemination of flames
(see Figure 9).

8(a) A Palmetto palm-Caribbean pine savanna in Belize. Pines are well adapted to the infertile soils and palmettos flourish where roots can access water;

(b)

8(b) **The lignotubers of the Palmetto form a root mass around ½ m long with associated surface and subsurface lateral roots penetrating the soil for up to 20 m.**

Savanna fires are the epitome of unpredictability. Their behaviour is notoriously fickle and, depending upon wind strength and direction, they can scorch across the landscape at truly frightening speeds or at times appear to get stuck and burn out. Their twists and turns have defied reliable forecasting until relatively recently. The movements and intensities are much better understood with the aid of remote sensing (RS) techniques from satellites, giving a spatial picture at high resolution and frequency (see Box 1). There has been an upsurge of interest on the nature of savanna fires and the biogeochemistry of the Earth's environmental systems. Examples include the NASA/University of Maryland Fire Information for Resource Management System, the Southern Africa Fire-Atmosphere Research Initiative (SAFARI), and the integrated fire management system of the Kapalga experiment in the Kakadu National Park of Australia. Considerable interest has been focused on the links between fires and emissions of gases and particles into the atmosphere, and, depending upon the wind systems prevailing at the time, the results of burning may be felt many thousands of miles away from the sources.

Most savanna regions suffer from fires at frequencies of between three and five years but in many areas annual burns are reported. The frequency and intensity of burning is a major influence in

9. The fissured bark of a fire-resistant tree in the Brazilian *cerrado*. The grooves are up to 2 cm deep.

Box 1 Remote sensing and savanna landscapes

Remote sensing (RS) can be defined as any observation made off the ground surface and consists of groups of techniques for mapping and monitoring. When illuminated by the Sun, all objects on the Earth's surface will reflect a percentage of the received radiation in varying amounts and wavelengths, and all objects emit radiation due to heat loss. Passive RS comprises techniques that can be used to collect the reflected and emitted radiation of objects using passive RS systems. In contrast, when illuminated by an artificial source, such as microwave radiation, objects may reflect or backscatter a percentage of the received radiation, and both radiations can be collected using active RS systems. The fundamental difference between active and passive RS is that the former comprises two components—a radiation emitter and a sensor; while the latter has only one component—the sensor. In both active and passive RS, the collected data are observed using sensors which are mounted on aircraft, drones, or satellites, and the data are represented as aerial photographs or digital images. Aircraft and, more recently, drones have been utilized where observations at a high resolution are required, while satellites have been used where a wider coverage but higher frequency is needed. However, satellite imagery is becoming available at increasingly higher resolution and the use of satellites offers the advantage of being able to monitor change continuously.

RS has generated data on the dynamics and structure of vegetation, land use, plant and animal diversity, gas emissions, biomass, and carbon dynamics, and supplied valuable information for modelling climate and wildlife change. Since scale is a crucial aspect of the varied nature of savannas, RS technologies have had to come up with ways of combining the broad-scale global images with multiscale, finer levels of analysis, but the identification of individual savanna species is difficult and expensive.

(Continued)

Box 1 Continued

The difficulty of sensing what is happening at the ground surface has been partly overcome by the use of active microwave sensors such as radar, which are capable of collecting data through cloud cover or during darkness. A number of active microwave sensors have proved to be particularly helpful in understanding savannas. Lidar scans a target using a laser beam and measures reflected light, and can be used to profile the heights of vegetation canopies at the scale of individual vegetation associations. ALOS-PALSAR 1 and 2 are examples of land observation satellites that collect radar data at longer wavelengths than other active microwave sensors and can be used to map above-ground biomass or to assess differences in vegetation structures. AVHRR gives a large-scale picture of the Earth with a time interval of twelve hours and a resolution of 1.09 km. MODIS is a moderate resolution spectroradiometer covering the entire Earth's surface every two to three days and has been especially helpful in tracking fires. Satellite sensors which can collect data at multiple wavelength bands (multispectral sensors), such as LANDSAT Thematic Mapper or SPOT have been used for decades to depict vegetation, track fires, and monitor land cover changes at ground resolutions from 30 to 5 m. For finer-scale mapping, the commercial IKONOS and QuickBird multispectral sensors provided imagery with resolution from 1 to 4 m from which tree crowns may be delimited, while more recent satellite sensors such as Worldview II and III provide panchromatic imagery at sub-metre levels, from which it is sometimes possible to identify individual trees.

Remotely sensed data are frequently downloaded into image-processing software for preliminary steps such as geometric rectification and enhancement. The analysis of images, such as semi-automatic and automatic image classification, generates data that can then be exported into Geographical

Information Systems (GIS). Here they can be combined with other types of digital geographic information to carry out tasks such as mapping and spatial decision-making.

Data acquired from RS also provide the basis of many types of mathematical modelling, aimed at understanding the nature and changes in savanna systems. Such data have been used in assessing, for example, the dynamics of disturbances such as fire or land cover clearance; to shed light on tree versus grass competition; on determining the role of moisture variations, drought stresses, and competition between C3 and C4 herbaceous plants; or measuring ecosystem services. Nevertheless, even when data are carefully checked by ground-truthing, which is still essential, there are inherent limitations in using radiated signals collected by RS techniques as a means of understanding life on the Earth's surface and the properties of materials located beneath the ground surface.

destroying or influencing the vegetation. Conversely, a period without burning can result in a re-growth of the plant life with a trend towards more woody vegetation and eventually a return to a wooded landscape. This has been documented in many parts of the world, and illustrates the constant struggle between the forces of destruction and growth. Fires can result in dramatic changes in the vegetation over short distances, and frequently result in sharp boundaries between burnt and relatively unscathed vegetation. These divisions persist after the end of the burning producing a patchwork appearance from the air, with plants regenerating at different time intervals and different stages in the theoretical progression towards equilibrium with climate.

As a result of prolonged burning, the vegetation structure is strongly affected by reducing the proportions of trees to shrubs and herbaceous plants, and increasing the numbers of annuals at the expense of perennials at the ground surface. Tree height diminishes and the above-ground biomass is markedly reduced although, once the woody vegetation reaches around 3 m or more, it becomes less susceptible to fire. Burning has little influence below ground (5–10 cm), except for fires of extraordinary intensity, but does modify the processes of nutrient cycling and affects gas exchange with the atmosphere. Burning releases nutrients from deep-rooted vegetation and can therefore act as a periodic nutrient-pump. The herbaceous stratum provides around 90 per cent of the fuel for fires, which have been recorded to reach 840°C at 60 cm above the ground, but the fires tend to be short-lived. Where savanna merges with a different type of vegetation, the behaviour of fires can change. At the more arid margins, the increasing sparsity of vegetation means that fire cannot traverse the landscape with the same ease that it does with continuous fuel-providing vegetation cover.

There have been numerous attempts to quantify the nature of fires either from observations of natural events in the field or by establishing long-term fire experiments. A number of well known long-term fire experiments were established in Africa during the colonial period and notably after World War II, and measurements have in places continued for fifty years or more. The Kruger integrated fire management system in South Africa, measuring both controlled and spontaneous burns, is amongst the best known, but controlled experiments have taken place in West Africa; the savanna woodlands in Zambia and Zimbabwe; along with very detailed observations in northern Australia; and smaller but valuable trials in South America, mostly located in Brazil or Venezuela. Many of these controlled fire experiments were based on surveyed plots and whilst they provide valuable indications of trends, they have been criticized for their rigid design and for not being an accurate mimicry of the uncontrolled fires in the wild. Inevitably fire experiments try to simplify the conditions being measured as opposed to replicating accurately the heterogeneous and uncontrolled natural environment. However, it is much more difficult to control the parameters of wild fires and so both approaches offer useful information.

10. Results from long-term fire experiments in *miombo* woodland, Zimbabwe: (a) grassland resulting from fifty years of annual burns; and (b) the control plot with no fires showing the dominant Brachystegia and Julbernardia trees.

All the studies agree that increasing the frequency and intensity of fires reduces the woody component of savanna and increases the proportion of grasses and herbaceous plants. Conversely, if burning is prevented, there is a gradual colonization of open savanna by shrubs and trees (known as bush encroachment in southern Africa). Plant composition can at times be surprisingly unaffected although there is an increase in fire-adapted species with continuous fire impact. For example fire experiments in the *miombo* have shown increases in specific fire-tolerant grasses and sedges (see Figure 10). Early ideas considered fire to be entirely harmful but fire models have been developed to understand the processes better and to offer strategies for control. Over the past few decades it has been shown that fire management with regular low-intensity burns can control woody encroachment, reduce pest infestation, and manage the fuel load more effectively, whilst encouraging nutrient return to the soil and at the same time enhancing plant diversity.

Beneath the ground—the influence of soils

The part played by the underlying nature of the soil is often overlooked in assessing the nature of savanna landscapes. Many aspects of climate, particularly rainfall and temperature, directly influence soil properties, and they, in turn, influence the vegetation and wildlife. The amount of moisture affects weathering and solubility of nutrients, while temperature influences the rates of organic decomposition and biochemical processes. There are many instances where the character of the soil overrides other determining factors in shaping the landscape. Parent material derived from the geology can affect soil properties to the extent that typical savanna plants are less competitive (as in calcareous and base-rich parent materials). Shallow, rocky soils such as those found on inselbergs, and kopjes in Africa, have limited rooting depth for plants (see Figure 11) although most savannas have deep, highly weathered profiles. Seasonally flooded areas also limit the characteristic savanna

11. A resistant rock outcrop (or kopje) in the grassland savannas of the Serengeti National Park.

shrubs and trees, which prefer good drainage. Islands and mosaics of differing plant communities are frequently a reflection of the changing nature of the subsurface and are not necessarily obvious from the ground.

Soils and plants evolve together but many soil profiles are much often older than the vegetation they currently support. Soil physical and chemical properties are also strongly affected by the structure and composition of the underlying rocks, but the surface soil characteristics are the horizons most in tune with the contemporary plant cover. Soils are often thought of as inert features of the environment, but in fact they are cauldrons of biological activity, and this is most clearly manifest in the surface horizons where they are in closest contact with plant and animal life. In the top 10 cm or so of the ground surface, the organic debris raining down from the vegetation is fragmented, consumed, decomposed, and redistributed by a host of organisms from small burrowing animals such as rodents and snakes, macrofauna (such as earthworms, termites, ants, and beetles), mesofauna (0.2–2 mm body size) such as mites, collembolans, and some nematodes, and millions of micro-organisms. Soil organisms act as decomposers (of organic matter), transformers (nutrients), engineers (soil structure), and controllers (of biological populations).

The availability of water for plants is closely related to topography and drainage. The physical shape of the landscape plays a role in drainage by shedding rainfall more rapidly over more steeply sloping ground and restricting lateral flow over flatter land. Convex upper slopes and concave lower slopes generate differing processes of erosion and deposition, producing a linked sequence or *catena* of soils (see Figure 12). Even slight differences in topography can affect soil drainage and plant response. Furthermore, at drier times of the year the surface soils dry out faster than the subsurface and consequently drainage is related to the permeability and texture of the whole soil profile. As rainfall

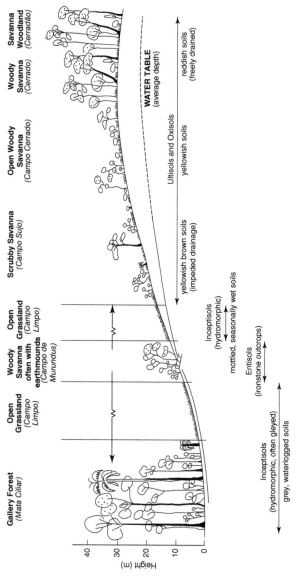

12. **The influence of slope and drainage on plant communities and soils over a typical catena in Fazenda Agua Limpa, Federal District, central Brazil.**

39

drains it leads to the mechanical washing, and *leaching* (chemical removal), of valuable plant nutrients. These are translocated either to lower positions in the profile or out of the plant rooting zone entirely. The presence or absence of a temporary or relatively permanent water table will also influence the ability of savanna plants to regulate their water supplies. Surplus water flows off the ground surface where the infiltration capacity of the soil is exceeded, as often occurs in the wet season, and may cause erosion as well as representing a potential loss of water to plants. The moisture storage ability of the soil is therefore a critical aspect of plant survival and plants have devised a number of ingenious adaptations to overcome the particular set of stresses that they face.

The time over which the landscape has developed is also significant. Many savanna soils over the ancient Gondwana blocks have developed over extremely old weathered land surfaces. Soil evolution leads to the formation of profiles with characteristic properties. As a broad generalization, the greater the extent of weathering associated with immensely long periods of time, the greater the degree of chemical and physical evolution (Oxisols and Ultisols in the American United States Department of Agriculture (USDA) taxonomic system; mostly Ferralsols or Acrisols in the Food and Agriculture Organization (FAO)/United Nations Education, Scientific and Cultural Organization (UNESCO) classification). These deep, acidic, infertile soils dominate the world's humid tropics, and the heavier rainfall regimes in South American savannas result in a far larger proportion of such soils, with lesser proportions in most of Africa and in the driest conditions found in Australia.

On the other hand, where the soils are less evolved or where a drier climate results in less weathering, the profiles retain many of the nutrients and are consequently more fertile (USDA: Alfisols; FAO: Nitosol, Luvisols). Over more recent savanna landscapes, such as alluvial plains, recently exposed marine terraces, or eroded rocky plateaus, the soils reflect the younger materials that have been deposited or exposed rather than the passage of time (USDA: Entisols and Inceptisols/FAO: a range of groups). In places, the influence of underlying geological parent materials is so strong that the soils reflect the original mineral constituents even after considerable weathering. Examples include the fertile soils of volcanic regions with base-rich lava flows (USDA: Andisols/FAO: Andosols). Base-rich and clay-rich soils are often dark coloured with distinctive properties such as those in the moister *dambos* or *vlei* in Africa (USDA and FAO: Vertisols). Where there are exceptionally coarse-textured and sandy soils, the soils become highly leached, giving rise to very acidic and infertile surface soils, and sometimes strikingly different subsoil accumulations of organic and mineral compounds (USDA: Spodosols/FAO: Podzols).

In seasonally very wet or *hyper-seasonal* areas, undecomposed organic matter accumulates to give distinctively dark-coloured soils (USDA and FAO: Histosols). However, in arid savanna regions, the lack of rainfall often results in soils that are hardly weathered at all and consequently retain many of the mineral characteristics they inherited from the parent rocks beneath (USDA: Aridisols/FAO: Solonchak, Solonetz). In these marginal savanna regions, the longer dry seasons can bake the ground surface, and the lack of rain combined with capillarity brings soluble salts to the surface, producing soils which are potentially fertile but lack moisture to maintain a continuous vegetative cover.

While the nature of savanna soils varies with the amount of rainfall and the age of the land surface, events at any one spot are likely to have been shaped by the actions of the current vegetation and its associated organisms. Moist savannas suffer from infertile soils with poor reserves of nutrients such as the exchangeable cations (particularly calcium, magnesium, potassium), low pH levels (indicating a greater acidity), very low available phosphorus (a key plant nutrient), and are often associated with weak organic matter levels (which act as stores for nutrients). At the same time, in extreme conditions, the high acidity is partly a result of high levels of exchangeable aluminium, which can become toxic. The remarkable adaptation of some plants in Brazil to these extreme conditions illustrates the way in which plants can develop adaptations over time in the face of extreme obstacles in the environment. Introduced crops (or exotic plants) find it difficult to compete and require careful management.

Animals and microbes

Living organisms shape their environment and savannas are fashioned by billions of microscopic creatures as much as by the much smaller number of large and visually dominant animals. At an intermediate scale of size, insects play a crucial role in grazing, pollinating, and distributing plant materials while forming a vital part of the food chain. Wildlife is controlled to a large extent by the nature of vegetation but equally it impacts on plant life. The nutritional value of the grasses, shrubs, and trees clearly affects the numbers of browsers, grazers, and their predators, but equally their selective consumption of vegetative material at different heights and seasons helps to determine the floristic composition that is best adapted. At the ground surface and below there are other organisms at work collecting, distributing, and breaking down organic matter which in turn influences the potential for plant growth.

The influence of animal activities in shaping the savanna landscape may be approached by considering the largest animals first. Some of the largest terrestrial animals on Earth flourish in savannas. They may browse the uppermost tree canopies (like giraffes) without causing much physical disturbance, or they may (like elephants) radically affect the vegetation by tearing down trees and shrubs, and opening corridors into wooded areas in search of food. With smaller sizes but often huge numbers, the grazers are characteristic animals of all the savannas with the exception of the New World, where they are present in much smaller numbers. Browsers can be broadly defined as animals that consume woody vegetation, whereas grazers focus on grass and other herbaceous plants. Grasses are well adapted to grazing and can regenerate rapidly. In Africa the savannas are typified by the herds of wildebeest, antelope, and other grazers whose annual migrations in search of food and different patterns of feeding affect the plant composition of the great savanna plains, a role to some extent comparable to that of the kangaroo and wallaby populations in Australia. In South and Central America, the large herbivores are replaced by smaller animals particularly deer and some of the largest rodents in the world (such as the capybaras) with a rich array of birdlife.

Smaller animals are also vital in affecting the landscape. Termites are sometimes referred to as the major engineers of savannas, since, over a long period of time, their burrowing and organic matter redistribution can have a startling visual effect as well as a less obvious subterranean influence. Termites form a distinctive feature of the landscape and they are familiar sights in most savannas. There are known to be over fifty species in the Order Isoptera living in tree nests, hollowed out cavities in woody vegetation, entirely underground, and visibly in mounds. The mounds contain a convoluted network of galleries and chambers that regulate the microclimate, especially temperature and light, and the surfaces are sealed from the atmosphere by saliva and excreta. Termite activity within the mounds can extend downwards into the soil by up to a metre or more. They can also construct towers several metres high (see Figure 13(a) and (b)), that develop a variety of shapes and sizes depending upon the species. Termite mounds also offer a habitat for a diverse range of other species. Different termite species occupy different sites within the savanna and have adapted to the environment in a remarkable number of ways, although their activities are not always benign since they selectively influence plant composition. Termites are particularly important for their ability to decompose grass, wood, and other organic materials. This results from the symbiotic links with protozoa and bacteria in the termite gut that permit the breakdown and digestion of cellulose and lignin.

Furthermore, the nitrogen-fixing bacteria enable atmospheric nitrogen to be fixed and utilized, and termites can therefore influence atmospheric carbon and nitrogen cycles. Fungi are also invaluable in decomposing wood for termite consumption (e.g. breaking down resistant lignins to simpler forms such as polysaccharides) and dispersing potentially toxic components. Studies have shown that nearly 40 cm of subsurface material could be translocated to the surface by mound-building termites over a period of some 1,000 years. Mounds can concentrate useful elements through the accumulation of organic and mineral matter, although the precise amounts vary considerably over different locations.

The micro-organic world is less well known but plays a vital role in determining the character of savannas. Examples from this enormous group of organisms include nitrogen-fixing bacteria, which form nodules on roots of many species of legumes and provide a vital nutrient in nutrient-poor environments; numerous fungi that exist symbiotically with plant root hyphae, helping to absorb

13(a) **Termite mounds in a eucalypt-grassland in northern Queensland;**

13(b) Section through a termite mound created or occupied by
Sinotermes, showing the intricate galleries reaching to a depth of
around 80 cm.

relatively immobile ions such as phosphorus, copper, and zinc; and saprophytic soil fungi that have a key role in decomposition processes mineralizing and recycling plant nutrients.

The factors that combine to shape world savanna regions vary over space and through time, but an understanding of the character of a savanna in any given locality is enriched by an appreciation of the determining processes. The compelling feature of contemporary savanna life, however, is the dominating effect of human activities. Human colonization of savannas has demonstrated an ability to destroy—through clearance of vegetation, reducing animal habitats, or by accentuating soil erosion and desiccation. At the same time, human ingenuity shows a capacity to enhance the landscape through conservation measures, appropriate fertilization, and moisture and pest control. Humans have become the principal determinants of the savanna landscape.

Chapter 3
Savanna vegetation

It might be imagined that as all the savanna regions of the world
have evolved through the same determining processes with similar
components, they would contrive to have a similar appearance.
But this impression would be far from accurate. Whether
savannas are natural or semi-natural, they have developed in an
uneasy equilibrium over time with the pressures outlined in
Chapter 2, leading to an amalgamation of landscape patterns.
Furthermore, many savannas are the product of disturbance and
conversion from other ecosystems, especially tropical forests. Each
continent has ended up with a unique collection of plants and
animals with a particular environmental evolution and history of
human occupation. As a result they have followed different
evolutionary pathways leading to distinct landscapes and
particular challenges for their futures.

Why do savannas differ?

As the southern continents moved northwards and separated
during the long coursing of continental drift, they lost and gained
animal species and moved into areas of greater or lesser rainfall. A
few characteristics of the old Gondwana blocks still persist, such
as the Proteaceae and Dipterocarp plant families with relatives in
Australia and southern Africa, or the Vochysia family found in
West Africa and in South America. However, these inheritances

diminished over the vast periods of time that have elapsed since the continental separations. Whereas many of the plants and animals have functional similarities (i.e. they occupy the same apparent niches in each biome and may have distinctive congruent features), they are no longer related at a species level. Thus while the vegetation appears similar because of the analogous environmental constraints, the plant composition and links with animal life have diverged. When the different continental histories of human occupation are added, it is not difficult to see why each savanna locality bears the imprint of different antecedents.

Primary productivity

A number of key aspects of vegetation underpin an understanding of the continental similarities and differences. At this stage it is useful to return to the issue of precise definition. Plants gain energy from the Sun by means of photosynthesis, which operates by different mechanisms or photosynthetic pathways. Most plants assimilate carbon (C) by means of a C3 photosynthetic pathway (the term given to forms of photosynthesis with a three-carbon molecule to capture carbon dioxide (CO_2). This mechanism prevails where there is sufficient moisture and moderate to cool temperatures with high CO_2 concentrations. However, savannas differ in that they have a dominance of C4 grasses (a photosynthetic pathway with a four-carbon molecule) without taking into account the number, density, or structure of the tree and shrub cover. The C4 pathway is an additional evolutionary adaptation, and allows plants to tolerate higher temperatures with lower water availability and frequently lower CO_2 concentrations. This physiological mechanism has been described as a CO_2-concentrating pump and most savannas are covered by grasses that have this pathway. However, the worldwide distribution covered by this definition is likely to underestimate what many people perceive as a savanna landscape. This is because there are some other open areas without a dominance of C4 grasses, for example: in more shaded and cooler conditions;

at higher altitudes with a risk of frost; in wet grasslands; or where there is an inclusive tract of forest. These differences are significant because the two pathways are associated with different grazing potentials; while C4 species frequently produce a greater biomass (total organic matter), the quality of fodder for grazing animals is lower than that of grasses with a C3 pathway. Strict definitions of savanna demand continuous or nearly continuous grass cover, but there are many areas with discontinuous or sparse grass dominance which still have every appearance of being a savanna landscape, particularly at the arid or forest margins.

The primary productivity or amount of energy fixed by plant life represents an important strand in understanding savanna vegetation. One of the most widely discussed topics at the present time, because of the growing interest in productivity of plant systems, is the amount and potential biomass found at different sites. *Gross primary production* (GPP) is the amount of chemical energy that primary producers create in a given length of time. Part of this fixed energy is used by primary producers for respiration and maintenance of existing tissues. *Net primary production* (NPP) is the remaining fixed energy (represented as biomass).

Savannas constitute the third largest world biome in terms of biomass production after tropical and temperate forests. As might be expected from the variations in vegetation and density of biomass, the different components of savanna contribute varying amounts to the total NPP. Taking the Brazilian *cerrado* depicted in Figure 12 as an example, a related example showed that the total above-ground biomass for all plant material ranged from 1.8 tonnes of carbon per hectare (tC ha^{-1}) in *campo limpo*, to 2.7 tC ha^{-1} for *campo sujo*, 9.6 for *campo cerrado*, and 11 for the dense arboreal *cerradão*. However, the subsurface organic matter is frequently higher than the above-ground biomass, and figures for the same site reflect the importance of the below-ground contribution, with

7.6, 15.1, 23, and 25.5 tC ha^{-1}, respectively. This provides a further instance of the need to take the subsurface into account when trying to understand the savanna landscape.

Vegetation structure and species richness

Savannas reveal surprising differences in floristic composition and species richness. The number of trees and woody species varies with the type of savanna. Taking the Brazilian *cerrado* as an illustration, the species count may reach over seventy in savanna woodland and drop to less than twenty in more open grassland (see Table 2). In ecological terms, this diversity is referred to as gamma (γ) diversity, representing a macro-scale view of the vegetation. Within a local scale such as a specific habitat, the plant variety is known as alpha (α) diversity, while the difference between different sites or habitats is beta (β) diversity.

Trees or grasses?

The scenic appearance of savanna clearly rests on the proportions of trees, shrubs, and grasses. The reasons for the dominance of one life form over another and the spatial variations that are found throughout the world often require a great deal of untangling. There are several theories as to how they co-exist. One explanation is 'niche separation', which implies that trees are favoured where their deeper roots are able to tap moisture at depths lower than the roots of most grasses and thereby occupy a different niche. One of the difficulties of this concept is that when the trees are young—at a seedling or sapling stage—they are probably competing at the same depths with non-woody vegetation and so both must co-exist at times. A second explanation is known as the 'bottleneck hypothesis', which suggests that trees would naturally colonize and later dominate the savanna but for disturbance events such as fire, drought, grazing pressure, or storms. These act as bottlenecks that restrict and reduce the density and cover of trees. The implication of this

Table 2 Characteristic vegetation properties of a central Brazilian *cerrado*

Vegetation properties	*Cerradão* (Savanna woodland)	*Cerrado* (Woody savanna)	*Campo cerrado* (Open woody savanna)	*Campo sujo* (Open shrubby savanna)	*Campo limpo* (Open grassland)
Canopy cover (%)	46 (15–85)	19 (1–55)	3 (0–15)	1 (0–3)	< 1
Height of stand (m)	9 (6–18)	6 (4–8)	4 (3–6)	3 (1–5)	1 (< 5)
Number of trees/ha	3,215 (1,643–4,925)	2,253 (836–3976)	1,408 (335–2,928)	849 (266–2,070)	< 266
Number of tree species/stand	55 (40–72)	43 (26–60)	36 (18–52)	31 (19–43)	< 31

Source: Table 1 in R.J.A. Goodland, A Physiognomic Analysis of the 'Cerrado' Vegetation of Central Brasil. *Journal of Ecology*, 59, 411–19 (1971). *Journal of Ecology* © 1971 British Ecological Society

idea is that trees will eventually outcompete grasses throughout the savannas unless or until there is disturbance. Some theories work in a number of instances but not in others, and so it is likely that a combination of ideas is required to account for the proportions of plants at any one place and at any one moment in time.

Much of the vegetation in savannas changes in appearance throughout the year. The changes in plant structure and function (or *phenological* changes) are associated with seasonal variations occasioned by temperature, rain, and day length differences. Consequently savanna plants have a great range of characteristics in addition to the broad categories of evergreen, semi-evergreen, and deciduous habit. The variations include the timing of seed germination, leaf appearance, pollination, flowering and fruiting, and stem and root growth. Some plants have seeds that resist light burning; some produce sharp thorns or unpalatable or poisonous chemicals to reduce grazing; a number of plants develop hairy, finely rolled, or hard leaves to reduce transpiration (or *scleromorphic* adaptation); some flower just before the onset of a wet season. In addition, plants have evolved techniques to deal with particular issues like nutrient deficiency, where nutrients are retained in the woody tissues before leaves are shed in the dry season.

Savanna scenery varies according to the density and regularity of the tree cover. Not only is there a spectrum from open grassland to savanna woodland, but there are also patches of forest within savanna areas and vice versa. This is particularly noticeable around the forest–savanna transition (or *ecotone*) and may be indicators of past climatic regimes, fire, or distinctive soil differences. Patches of different types of savanna vegetation are found throughout the landscape, for instance where there are rocky outcrops and shallow soils. On rocky uplands such as these, there are often dry, shrubby communities which can be rich in endemic plants as shown in the *campo rupestre* savannas of Brazil. At poorly drained sites, hyperseasonal or seasonally very wet communities prevail, such as the *veredas* of South America. Consequently the visual appearance of a savanna can range from uniform tracts of similar vegetation to a bewildering mosaic of different plant communities.

African savannas

Savanna is the most widely distributed biome. Around 40 per cent of the African continent is covered by savanna woodlands and a further 5 per cent by treeless grasslands but some estimates put the total savanna figure at 50 to 60 per cent (see Figure 14). The principal differences in climate compared with other savanna regions result from the continental shape, the location in relation to the Equator, and the distribution of mountains and elevated plateaus. These variations have given rise to distinct climatic and vegetation regions, from the well known east–west zones across the northern part of Africa, to the more diverse climates of the

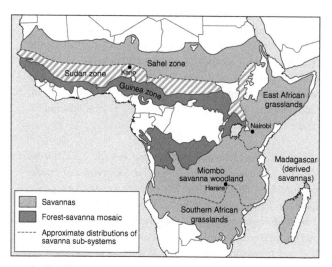

14. The distribution of African savannas and forest–savanna transitions.

mountain and plateau topography of regions through the eastern part of the continent. The ITCZ is also affected by the size and shape of the continent, extending to 20° north in July and 15° south in January. At the Equator, there are typically two rainy seasons over lowland areas because the sun passes overhead twice and these can merge to give more continuous rain with more monsoonal regimes. As might be expected, the rainfall becomes more variable and erratic as the climate becomes more arid, and grasses become progressively smaller, less dense, and with thinner leaves. In addition to the more predictable climatic variations, there are periodic disturbances that have a profound impact. Of these the ENSO events are outstanding, causing both flooding and drought over different savanna regions in recent years.

A number of distinctive zones have been recognized across the continent. The most extensive zone consists of *deciduous savanna woodland that* stretches 1,600 km along the plateaus running down the eastern side of Africa away from the coasts, from Tanzania to Mozambique and westwards some 1,300 km to the Angolan-Namibian border, with similar formations across West Africa. With drier conditions, a characteristically more *open and diverse savanna* is predominant and is most extensive north of the Equator. As the climate becomes more arid, the savannas merge into a *more semi-desert type* of vegetation varying from treeless, open plains to open woody formations and dense thickets.

Many of the grasslands are habitat-specific. Valleys and floodplains have better surface or subsurface water resources, and grasses predominate. This is because the sites are too wet for typical indicator trees such as acacias, and too dry in the dry season for any but drought-adapted woody species. Not surprisingly there is a great variety and abundance of grass species depending upon how much moisture is available and whether or not it is brackish. Typical wetter areas include the great swamps of the Okavango in Botswana and the floodplains of the Luanga River in Zambia or the Zambezi Valley. Some areas are markedly affected by the nature of the soils. For example, some regions are underlain by a hardpan, sometimes base-rich or alkaline, and this may result in short grass habitats as exemplified by eastern areas of the Serengeti, or the Makarikari (Botswana) and Etosha (Namibia) pans. Other areas have plants adapted to dark-coloured, dense, clay-rich soils (Vertisols), which are frequently intractable in the wet season and dry to brick hardness in the dry. Plants that survive these conditions include tall, coarse grasses and some drought-adapted acacias like the whistling thorn, *Acacia drepanolobium*.

West Africa. In West Africa the savannas cover an area of nearly 5 m km^2 and have been sub-divided into east–west trending zones. The most northerly zone is the *Sahel*, well known from its history of catastrophic droughts. The irregular rainfall is less than 600 mm per year with a dry season lasting seven to ten months. This zone stretches right across northern Africa south of the Sahara and north–south down the spine of eastern Africa from Somalia to the Kalahari sands that underlie much of Botswana, Namibia, and parts of South Africa. The typically thorny vegetation is dominated by acacias with only a limited number of other woody species occurring in areas with better water resources. Trees can still be found in the dry savannas but are confined to periodically watered depressions, streams, and subsurface moisture, but the overall appearance of the vegetation is of semi-desert steppe with drought-resistant bushes and annual grasses. The vegetation varies with the amount and distribution of rainfall. In the driest areas with yearly precipitation levels as low as 100–250 mm, grassland predominates with a woody component of around 10 per cent, made up of small trees less than 5 m high and scattered bushes. Acacias dominate and most of the grasses are annuals. There is an irregular and variable *ecotone* running to the desert margins, reflecting local factors such as soil properties, groundwater availability, and human activities.

South of the Sahel, the vegetation is commonly referred to as the *Sudan* zone. The annual rainfall lies in the range 500 to 1,000 mm with a dry season of five to seven months. The vegetation varies from *Isoberlinia* woodlands, with acacia more dominant in the north and characteristic wetter species in the south, to tall annual and perennial grasses. In the wetter areas to the south, the *Guinea* zone extends over an area of about 4,830 km east–west and 400–500 km north–south. Savanna woodlands flourish because of more favourable rainfall in the wet season, with the additional input from deep roots tapping reserves of soil moisture during the dry season. The more humid Guinea zone experiences over 1,000 mm rain per year, and in places double that amount, accompanied by a short dry season of between two and five months. The dry season typically runs from around December to February and is the period most affected by fire. Most woody plant growth occurs in the wet season but depending on the rainfall regime it is possible to encounter evergreen, semi-evergreen, and deciduous species. Most of the grasses, which may reach 3 m in height, and other herbaceous plants die down in the dry parts of the year, but rapidly regenerate and flower at the onset or even before the onset of rains.

Eastern Africa. The eastern savannas of Africa often provide the most iconic images of open grassy savannas teeming with wildlife (see Figure 15). They extend from the Horn of Africa to Tanzania ranging from a semi-arid to arid climate with seasonal rainfalls frequently below 500 mm to levels well over 1,000 mm. Much of the region experiences two rainy seasons related to monsoonal winds, within the periods March until May and from the middle of October to the end of December. The rains are notoriously unpredictable and unreliable. Temperatures on the other hand tend to remain constant averaging 25 to 30°C. Most of the region lies below elevations of 900 m and the landforms are extremely diverse, but the extreme weathering has resulted in generally infertile soils. You are never very far away from the influence of the Rift Valley, which generates a great variety of habitats.

15. Abundant wildlife in the East African savannas. The photograph shows mainly wildebeest and zebra in the Serengeti National Park, Tanzania.

The vegetation reflects this variety and includes many tracts that are not typically savanna, such as riverine vegetation, perennial grasses of damp depressions, and some deciduous woodland. Much of the woody vegetation is less than 5 m in height with characteristic genera of acacia and flowering *Grewia* (Tiliaiceae). The aromatic *Commiphora* (Burseriae) can reach 9 m in more favourable, moister areas. Plant life reflects the seasons and the uncertain rainfall but some plants persist throughout most years including perennial grasses and woody evergreens, the latter often forming thickets. A single resistant tree can provide a haven for shade plants and grazing animals beneath its cover. This process enriches the ground through the organic return to the soil, and encourages the outward spread of plants from these nuclei in the landscape thus forming small-scale ecosystems of their own. A number of the Rift Valley areas are affected by volcanic activity. This has produced some richer more alkaline soils, as in the southern tracts of the Serengeti, and the soil resources and catenas of the region are typically diverse, adding to the wealth of plants and animal life. The pattern of soils is further complicated by the widespread activities of mound-building termites that concentrate organic materials in their nests or by large burrowing animals such as warthogs or aardvarks.

Southern Africa. The *miombo* savanna woodlands are named from the Swahili word for one of the principal trees (*Brachystegia*). They cover an extensive area of nearly 2.7 m km^2, stretching from southern Tanzania to Zimbabwe and east–west from Angola to Mozambique, with outliers well beyond the present day core zone. The rainfall ranges between 650 and 1,400 mm in the wet season and *miombo* woodlands are sometimes divided into wet and dry, depending upon whether they have greater or less than 1,000 mm rain per year. This has a distinct effect on biomass with the wet areas nearly twice as productive as the dry.

Undisturbed woodland is typically 10–20 m in height with a nearly closed canopy of nitrogen-fixing leguminous trees, a fragmentary shrub layer, and a sparse but nearly continuous herbaceous stratum made up of grasses, sedges, and forbs. *Miombo* woodlands are estimated to contain over 300 tree species and are dominated by the family Fabaceae, typically made up of the tree genera *Brachystegia* (with seventeen species), *Julbernardia*, and *Isoberlinia* (see Figure 10). The grasses are mostly of the genera *Hyparrhenia*, *Loudetia*, *Andropogon, and Digitaria*. The trunks of trees are usually slender with a wide arching canopy, and there is a flush of vibrant red and gold leaves at the onset of the rainy season. In moist areas the trees remain in leaf for most of the year, but become progressively more and more deciduous towards drier areas. The characteristic *miombo* merges into *mopane* woodland (dominated by *Colophospermum mopane*), and dry but more fertile soils support acacia woodland in the river valleys of the Luangwa and Zambezi and towards the semi-arid margins. There are an estimated 8,500 higher plant species in *miombo* woodlands and over 50 per cent are endemic. *Miombo* soils are generally nutrient-poor and acidic, well-drained with low levels of organic matter. One of the ways that the plants are able to survive is through their association with an underground network of root fungi, which provides scarce nutrients. A large proportion of the woody species have deep tap roots reaching over 5 m, as well as dense branching laterals, and in this way protect themselves against the lack of moisture particularly in the dry season. Hill slopes influence the lateral drainage giving rise to typical catenary associations of plants and soils. Over the lower slopes for example, the moister, fertile soils support associations of the tropical tree *Combretum* and tall grasses, while treeless grasslands are found where the water table influences plant growth in valleys and depressions. Fires are widespread with return intervals typically two to four years in length. They are mostly started by human activities for hunting and clearing for pasture. As elsewhere in savanna, where burning and other forms of disturbance are excluded, the landscape returns to a dense wooded state.

Southern African savannas cover over 50 per cent of the landscapes and include some of the best-known national parks in the world, such as Kruger or Hwange. They stretch from extremely dry western formations in Namibia through Zimbabwe, the northern part of South Africa and southern Botswana, to the moister Mozambique savanna woodlands and the coasts of the south-eastern seaboard.

The key to the nature of the vegetation is once again the amount and distribution of the rainfall. The rains occur in the warmer summer months and the dry season lasts between two and eight months. Evaporation during the dry periods exceeds precipitation so water is always the limiting factor to plant growth. The rainfall ranges from 200 to over 1,000 mm per year, encompassing a dry season of over six months from April to October with a wide range of mean annual temperatures, including a potential for frosts at higher altitudes. It has been suggested on the basis of pollen records that the contemporary pattern of climate has possibly only persisted for some 1,000 years. The Quaternary Period witnessed changes from warmer phases coincident with savanna woodland, alternating with cooler phases favouring grassland.

The vegetation ranges from tall moist woodlands having an annual rainfall of around 1,800 mm to the sparse thorny bush grasslands at the arid margins with very low precipitation levels of around 50 mm in exceptionally dry years. Plant life has many features similar to other parts of Africa, with extensive grasslands and distinctive savanna woodlands, and tends to increase linearly in density and woodiness with rainfall up to around 600 mm per year. The varied plant communities include the so-called *bushveld* towards the southern Cape and shrublands at the western desert margins, the *miombo*, and *mopane* savanna woodlands to the north, and the floodplain grasslands of inland moist depressions and along watercourses. Much of the savanna occurs on the Kalahari sands, which are plains at around 1,000 m above sea level covering an area of 2.5 million hectares (m ha) and forming one of the largest continuous areas of uniform sandy soil in the world. The grassy formations continue to over 2,000 m in the eastern hills and contain numerous non-savanna tracts. The Kalahari Transect is a project designed to span the precipitation and vegetation gradient without the complication of differing soil types. This has provided valuable information on the ecology of the region, linking rainfall to plants (from < 200 mm per year in south-west Botswana to > 1,000 mm per year in western Zambia).

The southern African savannas have been classified into arid and moist ecosystems. This division is based on the length of the dry season. Dry savannas are defined as areas with over six months of dry season and less than around 650 mm rain per year. Arid vegetation formations are found over calcareous and other base rich soils, while the moist savannas tend to occur on non-calcareous, more acidic soils. The deeply incised river valleys such as those of the Zambezi, Limpopo, and Luanga receive less rainfall than the surrounding higher plateaux and frost is a regular feature of the higher interior regions. The distribution of these savannas is closely related to the base status of the soils. The diversity of habitats and plant communities can mask the basic division into moist and arid formations, which have been well described in the Nylsvley Nature Reserve in northern Transvaal. Here and throughout much of the savanna, the broad-leaved woody vegetation is found in the wetter areas (500 to over 1,000 mm rain per year), with infertile soils and characterized by trees such as the genera *Burkea*, *Brachystegia*, or *Julbernardia* and tuft-forming grasses. Drier more fertile areas (250 to 650 mm rain per year) derived from base-rich parent materials, have fine-leaved acacias and *Commiphora* trees with more continuous lawn-forming grasses.

Tropical America

Savannas occupy a substantial proportion of South America and are principally found in the woody formations of central Brazil

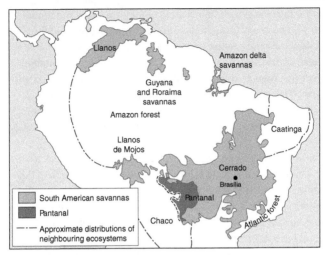

16. The distribution of South American savannas and associated ecosystems.

and the more open landscapes of Colombia and Venezuela (see Figure 16). Fingers of savanna vegetation extend northwards along the spine of the mountain ranges running through Central America with scattered lowland forms along the coasts and some remnants on Caribbean islands such as Cuba. The largest area is the Brazilian savanna, located mostly on one of the pre-Cambrian Gondwana plateau blocks, with extensions into the extremes of seasonally flooded or very dry soils around the peripheries. The savannas of the area north of the Amazon (*llanos*) present a different picture since they mostly comprise open grasslands that developed in more recent geological time (late Tertiary and Quaternary).

As might be expected, the climates over such an extensive region vary considerably and are affected by the shape and structure of the landmass. The greatest east–west dimension is found around the latitude of Brasília and land becomes constricted northwards in the tortuous isthmus joining the northern and southern continents. These differences between land and sea, highlands and lowlands, with different patterns north and south of the Equator, together with the spread of the area over many lines of latitude, mean that no one generalization about the determining role of climate will fit all locations. The essential features in common with the world pattern of savannas are warm climates and strongly seasonal rainfall, affected by movements of the ITZC and periodically dramatically influenced by ENSO events.

Brazilian savannas. The term *cerrado* is commonly used to describe these savannas, and literally means 'closed' (difficult to traverse on horseback), and implying a dense cover of woody vegetation. However, the *cerrado* is more widely understood to consist of a repeated pattern of grassland, and open and dense woodland, and originally made up nearly 25 per cent of Brazil. There are small extensions in eastern Bolivia and isolated outliers in the Amazon Basin and its surrounding countries, notably in the north-east of Brazil in Roraima and Guyana. The core area of the *cerrado* stretches from the borders of the Amazon forest south to São Paolo, west to the vast wetland of the Pantanal, and from sea level to over 1,800 m. Most of the characteristic savanna landscape is found over the plateau of central Brazil which acts as a divide for three of the main rivers of the country—the Amazon, the Paraná, and the São Francisco. To the north-west, the *cerrados* merge into the adjacent very dry woody vegetation of the *caatinga* and to the south-west into dry formations known as the *chaco*. Seasonal rainfall is clearly the dominant causal factor and the present climatic regime would encourage forest growth throughout the region, but it is taking time for the forest to spread. For example, some large areas, such as the Araguaia Basin, were once covered in *cerrado* and are now forested. Rainfall can vary between 800 and around 2,000 mm per year with a dry season running from approximately April to September, and a range of annual temperatures from 18 to 28°C. In some areas the onset of the wet season can be very irregular, and the dry vegetation is particularly vulnerable to fire, yet many plants begin their growth cycle in anticipation of the rains because they have deep rooting systems that can tap the water table. Frost can also be an inhibiting factor in the highest parts of the central plateau.

The *cerrado* represents an ancient vegetation formation. The earliest available records indicate that grasses had evolved at least thirty-two thousand years ago (Kya) over the central plateau, although it is likely that they are considerably older. The evidence for this comes from cores containing charcoal and pollen remains illustrating changes in plant cover through time. The appearance of the present day landscape possibly only dates back seven to ten Kya or less. Although the *cerrado* vegetation is predominantly made up of small but dense woody trees and shrubs, the region is dissected by large rivers and myriad small streams and depressions which often support dense gallery forest (see Figure 17). Over the centre of the Brazil there is usually a repeated pattern of vegetation related to slope, drainage, and soil (as in Figure 12).

Whilst in any one site there may not be an extensive list of plant species (α diversity), an adjacent locality is likely to contain many different species (giving high β diversity). When this is extended over the entire *cerrado* region, the accumulated floristic inventory is impressive and accounts for the richness of the biome, one of the biological hotspots of the world. Estimates suggest that there are of over 12,000 vascular plant species, or plants with tissues for conducting water and minerals (the xylem) and other tissues for conducting the products of photosynthesis (the phloem). Possibly 44 per cent of the plants exist nowhere else on earth. The woody species are generally short in stature with open canopies. They frequently have a contorted appearance resulting from the spontaneous sprouting of fire-affected branches, with thick, corky barks offering a protection against burning (see Figures 6(b) and 9). The leaves are mostly hard and scleromorphic with thick cuticles and hidden stomata. The two characteristic indicator

17. Aerial view of gallery forest threading through open savanna in Roraima, Brazil. The elliptical ground formations leading into the stream are probably the result of differential drainage forming earthmounds (or *murundus*) in some locations.

species are the sandpaper tree, a genus with one unique species (*Curatella americana*), named after its rough leaf surfaces, and a genus with 135 species of trees and shrubs (*Byrsonima crassifolia*). Both species are drought-tolerant, recover rapidly after fire, and are adapted to low nutrient availability. They also cope with acidity, frequently a high clay percentage in the subsurface, and poor drainage, and they are resistant to termite damage. Palms are widespread but not very diverse and there are small numbers of cacti and epiphytes. It is unusual to find a large area covered by one species but can be found occasionally—as in dense thickets of bamboo. Hemiparasites and saprophytes (or organisms that derive their energy from dead and decaying material), are also widespread. For each tree species there are some three to four non-tree plants, particularly in the legume, daisy, grass, and orchid families. In the herbaceous communities, annuals make up only around 5 per cent since the perennials can cope better with the shortage of water during the dry months. Plants have an array of other physiological and morphological mechanisms for dealing with drought and fire, including many species with underground tuberous roots which sprout rapidly when conditions improve. A number of woody plants possess virtually subterranean systems having vegetative reproduction (sprouting from subsurface structures) and with shoots interconnected below ground. A significant development over recent years has been the increase in invasive plants (possibly some 500 species) which are serious competitors for the indigenous plants.

The vegetation of the Brazilian savannas is closely related to the physical, chemical, and biological nature of the soils. The most widespread soil orders are Oxisol and Ultisol that can be generally described as soil profiles having good physical but poor chemical attributes, resulting from a long history of weathering and leaching. They are generally infertile soils, typified by acidity (low pH) and low levels of soil organic matter and plant nutrients especially phosphorus, calcium, and many micronutrients.

At the same time there can be high and sometimes toxic levels of elements such as iron and aluminium. Together with high concentrations of clay, these elements remain as resistant residues and give rise to the characteristic red-yellow soils colouring the landscape. Nevertheless the soils possess a firm structure with high aggregate stability and are therefore capable of being cultivated using modern machinery despite the nutrient limitations.

Fires are a feature of life in the *cerrados*. Evidence of fire has been found in the western state of Goias dating back to over 30,000 years whereas human occupation is generally believed to date back some 12,000 BP (Before Present). Consequently fire has raged across the savannas long before the intervention of man but human activities have increased the frequency of fire incidents as the region has been occupied. Most of the fires are surface phenomena consuming grasses and small leaves and stems. This is reported as forming around 97 per cent of the combustible load in the *campos* or grassland, dropping to 85 per cent in the woody savanna or *cerradão*. Since herbaceous material provides most of the fuel, flame heights range from less than 1 m to around 3 m and therefore do not affect the woody vegetation in many years and the canopy leaves are damaged but not destroyed. As shown earlier in other savanna areas, the vegetation recovers extraordinarily rapidly after fire, with ground vegetation appearing often within days and much of the herbaceous biomass replaced within a year. Plants whose surface organs may be affected can often be protected by having their incipient growth tissues below ground. Even 1–2 cm of soil acts as a useful protection and below 5 cm there is little impact from burning. Fires clear older vegetation and release valuable nutrients to the surface in the form of ash, so the occurrence of burning can have positive benefits. Indeed, some species appear to require burning to assist seeds to germinate. Overall, however, burning is likely to affect mortality rates, reducing biomass and increasing vegetative reproduction over other forms of regeneration.

Although the *cerrados* are often described as if they form a homogeneous region, there are important inliers and associated formations that add variety to the landscape. Linear threads of dark green, as seen from the air, show up the evergreen forest and wetland vegetation that follows the drainage patterns and can merge to form dense patches of forest where rivers and streams join. These gallery forests have probably formed important lines of connection for plants and animals between the core areas and the outlying savannas in the past. The larger tracts of wetland such as the Pantanal are not strictly savannas but may be classed as floodplain tree-grass complexes. However, in the dry season the grasses take on a dominant appearance and much of the region is effectively savanna. The poorly drained lower slopes and depressions in the savanna also contain greater numbers of moisture-loving plants of different genera. Some of these lower slope landscapes have a curious mound-like appearance. The mounds (known as *murundus*) are often the result of termites building up nests above seasonal flooding, but in other places the mounds are residual features resulting of water streaming down the slopes causing differential erosion. There are also substantial areas of deciduous forest that form over the better soils of limestone or calcareous rock outcrops, such as the spectacular wooded hills of Goias Velho to the west of the region. The central *cerrado* appears to represent the core of the South American biological diversity. As you move towards the Amazon outliers, such as Roraima in northern Brazil or in Guyana, the savannas become less floristically diverse. This trend continues northwards through Middle America.

The llanos. These distinctive flat or gently undulating landscapes stretch from east to west across the northern part of the continent from Colombia to Venezuela and the Guineas. The *llanos* occur over lowlands representing the infilling of an ancient geological downfolding of the rocks or geosynclines. They lie between the coastal mountains to the north and the Guyana Shield (forming the higher Gran Sabana) to the south. They are predominantly situated within the Orinoco River basin from 7 to 10° north. The savanna extends over 500,000 km² of low-lying plains (*llanos* signifying plains), generating a mosaic of open grassland with scattered shrub and tree communities dissected by gallery forest and in drier areas some tracts of deciduous forest. There are a few hills reaching 300 m, which are remnants of erosion, but most of the area lies below 250 m and is made up of alluvial, aeolian (wind-deposited), and high, older plains. The early European explorer Humboldt referred to them as 'monotonous steppes' but recent research has revealed a far more interesting and varied panorama currently threatened by intensification of land use.

The *llanos* are characterized by warm climates (averaging over 24°C) with strong seasonal contrasts and an annual rainfall of 800 to over 2,200 mm per year. The wet season runs from April/May to October when the prevalent north-east trade winds are replaced by oscillations of the ITCZ and heavy downpours frequently cause large areas to flood (see Figure 18). The flooded landscapes

18. A capybara family group at El Cedral in the *llanos* of Venezuela.

are similar to the Beni and Mamoré basins in Bolivia (Llanos de Mojos), which do however have a more interior or continental type of climate.

The differences in local climate and soil produce a more varied landscape than might be imagined from the generalized picture described earlier. The vegetation reflects this variety, with a heterogeneous mixture of life-forms from featureless grassland to bush islands and more woody domains where there are more favourable soil moisture regimes. Consequently there are constant changes in plant composition and structure across the region. Nevertheless, the same pattern of open grassland, shrubby grassland, woody grassland, and savanna woodland that has been described for the *cerrado* is also found in the mosaic of vegetation in the *llanos* although with fewer plant species. Despite this, the *llanos* are perhaps best characterized by their broad, grassland landscapes where the alluvial plains appear to be better suited to herbaceous plants than the *cerrado*. There is a more or less continuous grass cover 50–100 cm high, typically consisting of tussocky perennials such as the widespread *Trachypogon* genus, with numerous sedges. The neotropical woody yardstick of the savannas, *Curatella*, is found in the more shrub- and tree-covered areas. Most of the above-ground grass elements die off or are burnt during the dry season but different species have varying growth strategies, flowering, and seeding patterns. Seasonal swamps and hyperseasonal grassland frequently occurs in the wetter districts with the characteristic palm (*Mauritia flexuosa*) appearing along drainage lines with subsurface water. Seasonal flooding is widespread over the *llanos* and may persist for six months in places with water depths reaching 2 m. Pollen evidence suggests that the current vegetation dates from the last glacial (early Holocene) 9730 to 5260 BP, with varying drier and wetter phases and human influences become very evident after around 3600 BP. The generally high rates of photosynthesis are counterbalanced by poor nutrient availability and lack of water during the dry season. Today as in other forest-savanna transitions in South America, wetter climatic conditions seem to be favouring an increase in the woody components of the *llanos*.

The *llanos* have mostly evolved over highly leached soils derived from Tertiary sediments and more recent Quaternary alluvial deposits. Most of the area is underlain by acid soils of very low nutrient status but the hilltops deposits and dissected high plains have better soil fertility though greater vulnerability to drying. Some of the sandy plains to the west along the Meta River in Colombia have higher water availability and retain moisture more effectively.

Central America and the Caribbean. This extremely varied area consists of the main isthmus of Central America running from Panama in the south to Mexico in the north and also includes the Caribbean islands. There are isolated patches of savanna scattered throughout the region. Where they occur over around 1,000 m they are more akin to temperate regions of North America with a greater preponderance of C3 grasses reflecting the cooler conditions. Below around 1,000 m the savannas are moulded by topography, drainage, and soils as much as the determining seasonal climate. A further characteristic is the occurrence of periodic catastrophic storms and hurricanes which can have a devastating effect on the vegetation.

These differences are well illustrated by the savannas in Belize. The upland savannas found over the Maya Mountains lie at altitudes over around 500 m and are covered by Caribbean pine, oak (close to its southernmost range), and a range of herbaceous plants dominated by tussocks of C4 grasses such as the genus *Andropogon* (see Figure 19(a)). The uplands are the remains of an ancient (Palaeozoic) land mass that possibly protruded above many of the marine incursions of later periods and have evolved into a highly weathered, undulating landscape. The steeply dissected slopes favour rapid run-off and drainage over the elevated sections and this downwash accumulates along with eroded soil and nutrients over the footslopes, giving rise to

19(a) The landscape typical of the upland savannas at around 1,000 m over the Bald Hills in the Maya Mountains, Belize; (b) An eroded and highly weathered soil profile in the Maya Mountains of Belize. The image shows a stone line mostly derived from volcanic dykes that penetrate to the ground surface.

distinctive gallery forests and a characteristic *catena* pattern of soil properties and vegetation. The soils are extremely infertile and exhibit all the signs of long-continued, intensive leaching (see Figure 19(b)). The recurrence of fires and hurricanes had led to a cyclic pattern of destruction and regrowth so the landscape never seems to remain the same. The lowland savannas by contrast are close to the sea, as in many parts of Middle America, and are typically very flat and poorly drained. The poor drainage results in part from the presence of thick layers of subsurface clay. The soils are once again acidic and infertile, supporting a pine-grass savanna with numerous inclusions of wet-loving species such as sedges and the widespread distribution of palmetto palms (*Acoelorraphe wrightii*) and calabash (*Crescentia cujete*). The resulting savannas have proved very difficult to utilize from the days of the indigenous Maya to the present time. This combination of well-drained upper slopes and poorly drained lowland plains is typical of much of the isthmus and is repeated to varying degrees over a number of Caribbean islands.

Australasian savannas

This widespread region consists of a series of landscape terrains rather than a zone of continuous vegetation. The savanna occurs where the annual rainfall is 1,000 to 1,500 mm, with a marked dry season of five to seven months and typically poor soils (Oxisols). The intensity and variability of the dry season determines the vegetation cover. Savannas merge into adjacent ecosystems, varying from arid to the wet flooded Terai-Duar grasslands close to the Himalayas, or into dry grass scrubland created by clearance and fire. From India eastwards to Indonesia, the savannas tend to be patchy and dominated by dry deciduous woodland.

The wet grasslands to the south of the Himalaya from Bhutan to Nepal have some of the world's tallest grasses—elephant grass (*Saccharum* spp.) reaching 7 m. They are spread over riverine plains and are regularly flooded, providing rich habitats supporting a prolific but endangered wildlife. The Chitwan National Park in southern Nepal embraces everything from evergreen forest to stretches of open grassland. It is a remarkable remnant of what must have been an extensive zone, and still represents a haven of luxuriant plant growth and home to a profusion of wildlife, from very large animals to a host of small grazers and browsers, along with reptiles and a rich birdlife. The savanna woodlands of South-east Asia are characterized by trees of the Dipterocarp family and were formerly more widely ranging, probably spreading out during the drier phase of the Late Glacial period. Since then they have retreated to poorer, rocky or shallow soils. However, many areas classified as dry forest are characterized by fire-resistant trees within a predominantly C4-grassland understory, and are possibly better classified as moist savannas. A large proportion of the savanna woodlands are found at low elevations below 100 m above sea level, and tracts of open grass can still be found in wet coastal areas, but they have been severely disturbed by human activity. An example is the south-west of New Guinea where the environment and vegetation is similar to that of the northern Australian coast.

Australian savannas. Savannas occupy more than a quarter of Australia (some 1.9 million km²) and are mostly found in the north of the country, although very dry landscapes resembling true savannas are found over very large stretches of the country (see Figure 20). Consequently, there are many different savanna landscapes which broadly reflect the patterns of rainfall and soil properties. However, the generally dry climate makes Australia the driest of all the continents and particularly prone to fire. Australian savannas have been characterized by extreme infertility and highly variable climate.

The savannas have expanded relatively recently in Australia's long evolutionary journey, with warm wet and dry seasons developing most strongly over the past ten million years from earlier warmer and wetter conditions that favoured evergreen forest. Today the

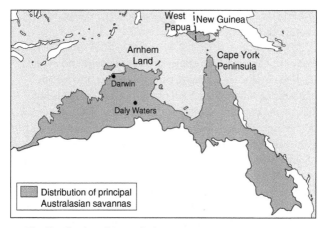

20. The distribution of Australasian savannas.

climate is typically monsoonal with hot and wet alternating with long, warm dry seasons. Rainfall relates closely to latitude, and the northern sections are moist with over 900 mm per year and in some places over 1,800 mm. However, the main savanna regions of Australia are semi-arid with annual rainfall levels of between 300 and 900 mm. From what is generally known as the 'top end' (Kimberley and Cape York Peninsula), rain declines southwards and also decreases from east to west away from the Queensland coast. A significant feature of the climate is its irregularity with the onset of the rainy period varying from year to year as well as the amount of precipitation. Generally the rainy season begins in September/October and runs until May/June when there is a period of rapid drying. The rainfall is generated from monsoonal troughs, which tend to have a stronger impact nearer to the coasts, and from isolated convective storms with added periodic downpours from cyclones. An influential feature on the climate is the variation in ENSO that results from changes in pressure between the Pacific and Indian Oceans. During an ENSO event the eastern regions tend to have a reduced rainfall while at other times there may be devastating floods, and extremes at either end of this spectrum can have a marked effect on vegetation structure and plant composition.

Most of the Australian savannas are found below 500 m and variation in relief is also slight, usually less than 100 m. The landforms and their related soil properties have a significant effect on the vegetation and these have been divided into three main groups. The *savanna woodlands* cover the level and slightly undulating landscapes making up most of the northern areas, although topographically steeper and moister to the east in the Great Dividing Ranges of Eastern Queensland. More *open savanna* covers the so called 'stone country' characterized by rocky escarpments and shallow soil spreads over the eastern part of Arnhem Land, the Victoria River in Northern territories, the Kimberley area of Western Australia, and the north-east of Queensland's Cape York peninsula. Along contemporary flood plains on Quaternary deposits, and some older sediments, there are *black soil plains* (named after the colour of the dense clays that develop cracks in dry periods). Throughout all of these subdivisions of the savanna landscape, there are sinuous stream lines with typically riparian vegetation that contrasts vividly with the desiccated vegetation of the dry season.

The vegetation ranges from treeless grasslands to open woodlands with a scattering of low trees in the arid interior covering less than 1 per cent of the area, to more dense woodland with a significant proportion of evergreen species closer to the rainforest margins (with up to 60–70 per cent coverage). Even in the dense savanna woodland, light intensities are high and there is fierce competition for water and nutrients. The woody species can range from evergreen to semi-deciduous to completely deciduous across the landscape. The most extensive grasslands are found in the more arid areas where there are the cracking clay soils, as on the Barkly Tablelands, with different grasses appearing as the climatic regime becomes moister. Grasses also predominate over the coastal lowlands.

Most of the vegetation is made up of eucalypt open woodland with acacias widespread over drier areas, but there are many variations at a local scale and numerous shrubs resulting from differences in rainfall and from changes in soil texture. Palms and cycads are also found throughout the savannas. The eucalypts can reach 10–20 m or more in the higher rainfall areas with canopy cover of 60 per cent, where *Melaleuca* spp. may be more dominant (a genus with most species endemic to Australia). In the drier savannas the trees are shorter (5–15 m tall) and the canopy cover is much less (5–30 per cent). In drier areas grasses and tough shrubs prevail, such as Aristide and spinifex, and these plants are usually resin-rich, which adds fuel to the spontaneous or deliberate fires.

The savanna plant and animal biodiversity is difficult to separate from the wetland ecosystems to the north, and from the scrublands and heathlands of parts of Kimberley and western Arnhem Land. This is because there is and has been continual crossover of species. Plants have evolved several mechanisms to cope with the dry periods including shedding leaves, sclerophytic adaptations, and an ability to continue photosynthesizing at times of water deficit. These characteristics are well illustrated by the huge variety of thorny shrubs like acacias (e.g. the mulga, *Acacia aneura*). Many plants have underground root tubers that allow rapid re-sprouting after burning, including trees such as the jarra (*Eucalyptus marginata*). Fire is an ever-present condition, preceding the arrival of the first inhabitants, through lightning strikes but augmented historically as humans cleared the vegetation for hunting, fuel, pest control, or to encourage plants that were more useful.

Many of the soils have weathered over long periods of time. They are generally infertile and plants need to conserve nutrients by protective devices such as nutrient removal from old leaves before litterfall. Mycorrhizae, the fungal connections below ground that connect with roots, become particularly important to the ecosystem in these circumstances, for example in assisting phosphorus uptake in eucalypts. Nearly all the savannas have soils that are turned over by countless generations of termites: their activities, together with savanna wildlife, are examined in Chapter 4.

Chapter 4
Wildlife and microbes

Savannas teem with life, but not all of the wildlife is immediately obvious. The lives of some savanna organisms can be tracked from their activities while others are hidden from view because of their size or habitat. Some may be identified by sound rather than by sight, such as many of the night-time animals. Wildlife depends upon the food resources provided by plants. These are the primary producers because they are able with the help of solar energy to convert inorganic materials into living matter. The primary converters of the Sun's energy in turn form the food resource for animals and micro-organisms and in this way they construct a food chain. Eventually all the living tissues become dead organic matter, and energy is released during the processes of decomposition.

Within the savanna landscape, myriad food chains make up a food web or energy flow of such complexity that research has only been able to mimic a portion of the total picture. Furthermore, the continents differ greatly in the numbers and visibility of the wildlife that lives in or utilizes the gamut of resources. This chapter looks at the pathways of energy through the savanna system and the nature of the wildlife that characterizes each of the major savanna landscapes. This brings in consideration of the food chains and food webs as well as aspects of animal ecology that determine the number and character of living organisms.

Energy pathways

Energy is the driving force behind all life and can be visualized as a constant flow from the Sun to the Earth. Only around half of the solar radiation reaches the Earth's surface because of absorption in and reflection from the atmosphere. The amount varies with latitude, height above sea level, and, at any one location, the weather conditions such as cloud cover. The numbers and activities of animals and micro-organisms depend upon the primary producers, and this is most easily portrayed as a flow of energy. Energy is transferred in successive stages (or *trophic* levels) from primary plants or *autotrophs*, to the secondary consumers or *heterotrophs* (meaning that they can gain nourishment from a variety of sources). These fundamental concepts are demonstrated in an example taken from one of the well-researched studies, the savanna woodland area of Nylsvley in South Africa (see Figure 21).

The Nylsvley study illustrates several useful ideas in understanding savanna wildlife. The primary producers at the first trophic level fix energy used in respiration, growth, and reproduction. Some

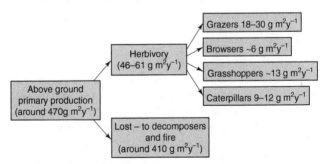

21. **A simplified food web at Nylsvley in South Africa. This illustrates the large proportion of the primary production that is lost to the food chain. There are very large yearly variations in production and in the levels of consumption.**

of this energy is passed on to secondary consumers like animals and insects but a significant proportion is unused and therefore passes on to various decomposition food chains and is ultimately dispersed as heat. The numbers at the higher trophic levels depend upon markedly larger numbers lower down the food chain and this is the main factor limiting the number of animals. It is often referred to as a pyramid of numbers or biomass, and the ecological efficiency in trapping and utilizing the initial energy input can vary enormously (less than 5 per cent to around 60 per cent). Since each stage in a food chain loses energy, mostly as heat, there are a limited number of possible levels—normally no more than five. However, energy is not the only factor determining the length of the food chain since the size of organisms, the nature of the predator–prey relationships, and the social structure (e.g. whether animals share resources) all play a part in shaping the character of exceedingly complex food webs.

Wildlife populations and communities

The life cycle of all living things in the savanna, as elsewhere in the world, depends upon a multitude of inorganic and biotic pressures. The balance of plants, animals, and microbes is constantly changing in an ever-evolving adaptation. Wildlife is continuously on the move and therefore the processes of dispersal, colonization of new territories, and reproduction change all the time, despite the recurrence of periodic or seasonal cycles such as the great migrations of wildebeest in the Serengeti plains of East Africa. The density of animal life is one of the wonders of the grassland savannas and depends on a variety of factors such as food supply, disease, and competition, as well as environmental stresses. The total population or number of a particular species, and its distribution, is a necessary basis for estimating change over time. Inevitably, it simplifies the complete picture of interacting communities of living organisms. Whilst the numbers of large visible animals can be estimated with reasonable accuracy, the numbers of smaller creatures and the ways in which they live together is a much more difficult undertaking, and many of the components of intricate food webs are far from understood. The two main aspects of wildlife to consider are the individual populations of animal species and the way they integrate into communities.

Populations. One of the major processes influencing the numbers of animals is referred to as 'density dependence'. As the name suggests, this happens where population growth rates of a species are regulated by their densities. For example, an organism would find the probability of finding a mate very difficult if the population densities were low, but would have a higher likelihood with higher densities. Such a situation is described as positive density dependence. Conversely, there would be a negative situation where population growth was limited by density, for instance where there was crowding, competition, or predation. Population growth is also affected by the numbers of offspring or the quality of care given to offspring. The differing reproductive strategies are given the terms r and k, where r selection occurs when an organism has a high rate of reproduction with low attention to care (i.e. quantity). Most insects in the savanna come within this category. On the other hand, k selection describes situations where there is a high relative cost of reproduction leading to a smaller number of offspring (i.e. quality), and elephant populations are good illustrations of this.

Communities. The interconnected nature of wildlife reflects animal–plant–microbe interrelationships and savannas contain many fascinating examples of how the interplay of these organisms affects the visual landscape as well as the invisible world below-ground. The savanna community is greater than the summation of its constituent parts and the cumulative impact on the landscape is what is seen by an observer.

The food web briefly outlined earlier expands into a vast network of interactions at any one locality. Some of the best examples have been worked out over many years in African national parks such as the Serengeti-Mara (see Figure 22). Although there have been numerous critiques, the general concept behind food webs and diversity is that the more complex the communities with greater numbers of species, the more stable will be the ecosystem. This is exceptionally difficult to prove but there are many signs that it does work and this has considerable significance for management. There have been many mathematical simulations or models trying to calculate what would happen given different scenarios. One group of models has examined the impact of disease on wildlife. An interesting example has been the viral disease rinderpest which badly affected cattle in Africa. This devastating disease afflicted cattle and ungulate populations for years after being introduced probably from Asia. Fortunately, the disease was eliminated officially in 2011, and subsequent monitoring in the Serengeti has shown how the wildebeest and giraffe populations have increased. This has had an increased grazing and browsing effect, leading to tree recovery and increased carbon stored in biomass.

Biological interactions

Life in all savannas depends upon an almost unimaginable complexity of function and behaviour between the smallest and the largest organisms. These species' interactions take many different forms. Where there is a predator–prey association, one species clearly gains at the expense of another, as in the cases of the lions and the wildebeest populations in Africa, and parasites and their hosts. In other cases, interaction between organisms occurs but is not fatally detrimental to either—as in many forms of competition. Mutualism occurs where two organisms depend on each other and both gain from the relationship. Even this simplified picture is not the complete story because each organism harbours a host of microbes which in turn may be antagonistic,

Wildlife and microbes

83

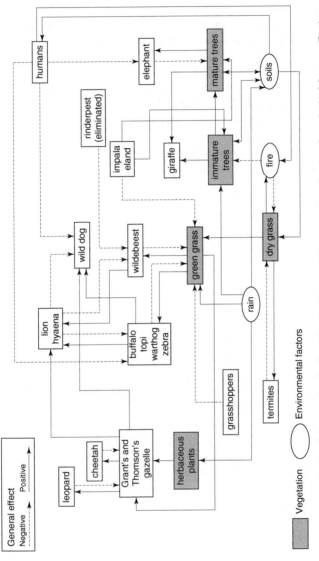

22. A representation of the complex food web in the Serengeti-Mara savanna, showing the principal factors affecting the interactions between the main animal components of the ecosystem.

General effect
Negative ←→ Positive

⬭ Environmental factors

▨ Vegetation

84

helpful, or neutral. However, it is not necessary to explore every facet of the interconnection to appreciate how they affect the savanna landscape any more than it is required to understand all the contributors to a food web to get the broad picture. A selection of examples serve to illustrate how plants and animals depend on one another and helps to visualize their role in maintaining the fascination saga of savanna life.

Ants and acacias. One of the classic examples of mutualism is found in Central American woody savannas. The bullhorn acacia (*Vachellia cornigera*) forms a symbiotic alliance with an ant species (*Pseudomyrmex ferruginea*). The ants utilize swollen hollowed-out thorns on the plant for shelter. The tree supplies food in the form of proteins from the tips of leaflets and carbohydrates from glands on the stalk. In return the plant gets a defence against herbivores (such as leaf-cutter ants and crickets), browsers, and grazers (including livestock and goats). The ants also keep the photosynthetic surfaces free of invasive weeds and even destroy seedlings of competitors from around the plant base. They further appear to keep harmful pathogens in check, such as bacteria (which also have symbiotic links with the ants). An analogous relationship is found elsewhere in the savannas. In Africa, the whistling thorn (*Acacia drepanolobium*) is home to several species of ants which drive off the smaller grazers. Not all of the association is necessarily beneficial, since the ants selectively affect the plants in and around the host. Figure 23 pictures an African example with *Pseudomyrex* ants colonizing *Acacia collinsii*. Yet there many large browsers, and it has been suggested that the process of browsing stimulates the development of swollen thorns and the production of leaf nectar. However, some other ant species are parasitic and it is not clear when and how these mutualisms evolved. What it does illustrate is the complex and convoluted nature of interconnected life.

Elephants and trees. Other forms of interaction are less benign for the plant. A good example is that of the elephant–tree relationship.

23. An example of mutualism in the savanna: the ant–bullhorn acacia symbiotic association.

It is difficult to imagine how the tree might benefit from being stripped of its leaves or completely uprooted, and it is estimated that a single elephant might uproot as many as 1,500 trees in a single year. The reduction of the Seronera woodlands in the Serengeti since the 1960s has been directly attributed to elephant damage, with subsequent regeneration affected by fire and by giraffe browsing. However, this form of interaction is part and parcel of daily life in the African and Asian savannas, and may be part of a continuous cycle oscillating between a mature woodland canopy and open grassland. From a broader perspective, there may be a value in opening up the tree cover for a range of plants that demand more light. From a pastoralist's point of view, for instance, this would be a benefit in limiting 'bush encroachment' on grassland.

Plants and fungi. A less visible but extremely important component of the savanna is that of subsurface fungi associated with plant roots (or *mycorrhizae*). Both organisms benefit from this ancient relationship going back to the earliest plants and they provide a good example of mutualism. It is estimated that perhaps 95 per cent of world plant species are linked with mycorrhizae, and in the majority of cases plants could not survive without them. The fungi live in and around the roots of savanna trees and shrubs. In return for carbohydrates supplied by the plant from the products of photosynthesis, they provide valuable nutrients drawn from the soil. Some critical nutrients are in very short supply in the root zone of the native plants such as phosphorus and nitrogen, and the presence of mycorrhizae can remedy many of the deficits by extending under the surface over huge spaces in a continuous interlinked network. There are many different types of root fungi but the two main groups are known as endomycorrhizae (or arbuscular, named from the arbuscles or nodules on the highly branched hyphae) and ectomycorrhizae. The endomycorrhizal group make up around 85 to 90 per cent of world plants and the fungi penetrate into the root cells (endo = inside). The smaller ectomycorrhizal group (i.e. outside) are nevertheless very important in savanna soils where there are often severe nutrient deficiencies. The fungi do not penetrate the root cell, and the exchange of nutrients takes place through the cell walls with the hyphae forming a sheath around the root. The principal *miombo* trees described in Chapter 3 have a widespread network of root fungi interlinking individuals over large areas. Ectomycorrhizae are of widespread ecological and economic importance, and two of the main plant families found throughout the tropics—the Dipterocarpaceae in South-east Asia and the pan-tropical Caesalpinaceae—depend on these types of biological interaction.

Wildlife and landscape

Landscape is a concept that is particularly appropriate for an understanding of animal life. Animals and micro-organisms are so mobile that they cover every part of the terrain and many organisms depend upon resources that are not strictly defined as savanna but are contained within the wider concept of landscape. The landscape resembles a theatre across which animals and

microbes pass and live out their roles on a stage that is itself
slowly changing. Thus what we see today represents an inherited
past as well as the geography of the present day.

This section portrays the diversity of savanna wildlife by considering
the ecology and dynamic processes that operate in a representative
selection of sites within each of the continents.

Woody savannas of the cerrado. Although the *cerrado* contains
some of the most species-rich vegetation of any savanna region,
it lacks the diversity of large animals that can be found in many
parts of Africa. This partly results from the extinction of the
ancient South American megafauna and the introduction of
a new fauna following the opening up of the Central American
interconnecting bridge. Nevertheless the *cerrado* harbours a
multitude of smaller animals and invertebrates as well as a rich
bird life. Around 760 species are known or assumed to breed in
these woody savannas. They make up nearly 50 per cent of Brazil's
avifauna with perhaps thirty species that are endemic. The bird
populations are good illustrations of the difficulty of associating a
species with a particular habitat. Their mobility enables them to
cross and utilize many different habitats within and around the
savanna, notably the gallery forests and edges of the Amazonian
and Atlantic forest formations. Some humming birds, for
instance, may utilize the *cerrado* during the wet season but stay
closer to the gallery forest in the dry times of the year. Seasonality
is also a driving force affecting insect, fruit, and flower abundance.
The incidence of fire, a defining characteristic of savannas, also
affects animal behaviour. For example, a number of birds have
adapted to tracking fires in order to catch insects and other
arthropods. There are numerous small mammals ranging across
the *cerrado*, which is the most species-rich biome in Brazil after
the rainforest areas. The mammals are generally small with
around 85 per cent having a body mass under 5 kg and only five
species with a body weight of 50 kg and over. Around 194
mammal species from thirty families have been reported, with the

largest groups being bats and rodents. The latter carry out the role of grazers and some reach very large sizes, such as the capybara (*Hydrochaeris hydrocharis*). See Figure 18.

Despite the limited number of large animals, there are between ten and fifteen carnivores that live within or utilize the *cerrado*, and the patterns of predator–prey linkages are similar to those in the better known savannas in Africa. The mammals are generalists rather than specialists and range over the whole of the savanna, but they are locally quite rare and not often seen. As with the associated vegetation they typically show high beta (β) diversity. The animals have probably adapted from forest areas in the past and this may account for the low levels of endemism (around 18 per cent). The vast invertebrate populations are only known for limited parts of this very large area and yet contribute substantially to the diversity of life and the nature of the food chains. For example, the butterflies and moth species (Lepidoptera) appear at first to be present in a bewildering variety of locations, but they have been shown to be sensitive to humidity, and therefore concentrated in gallery forests and dense vegetation with access to water. The proportions of endemic, widely dispersed, and peripheral species vary widely amongst different groups. There are possibly less than 20 per cent endemic to the *cerrado* biome and about a third distributed over the continent. Surprisingly there are very few links with the dry woody peripheries, such as the *caatinga* (to the north-east) and *chaco* (to the south-west).

Although the *cerrado* is under considerable threat from development, the protected areas of central Brazil in the state of Goias have retained a multitude of animals in areas of outstanding natural beauty. Two of the best known are the Emas and the Chapada dos Veadeiros National Parks. The Emas reserve (named after the flightless bird) contains many fascinating animals such as the giant anteater (see Figure 24) and giant armadillo, the rare maned-wolf, jaguars, and pampas deer. Termite mounds are a

24. A giant anteater in the Pantanal, Mato Grosso, western Brazil.

feature of the landscape and can reach over 7 m in height and 30 m in diameter, and arboreal termite nests are widespread. Many of the birds are grassland specialists as a large proportion of the reserves is open savanna. However, both national parks are threatened by the pace of current development affecting species like the blue and yellow macaw, the dwarf tinamou, and the yellow-faced parrot. Although the Emas park covers some 126,000 hectares (ha), the entire area is now surrounded by farming, and there is the added threat of uncontrolled fires that not only rage across the open land but cut into the gallery forests. The scenery at the Chapada dos Veadeiros differs from that of the typical *cerrado* in having rocky canyons and numerous waterfalls in elevated stretches, backed by tabular hills as well as broad grassy plateau surfaces. Most of the animal life is similar to that of Emas and the two parks together are believed to hold around 60 per cent of the plant species and 80 per cent of the vertebrates described for the *cerrado* region. Consequently these areas have been designated world heritage sites as habitats characterizing this major expanse of savanna.

Grasslands of East Africa. The Serengeti is a vast, mostly treeless grassland set over rolling plains. The Ngorongoro Conservation Area (NCA) forms the eastern section distinguished by its escarpments and spectacular crater. The reserves have been designated as world heritage sites, both from their ecological value and also because of the rich history of Masai settlement. To the south the grasslands are bounded by the edge of the plains and to the north and west the savannas have largely been converted into agriculture. The Serengeti has been described the 'epitome of a wildlife spectacle'. It represents one of the most extraordinary places on Earth for observing very large numbers of large animals and their predators. Although it is dissected by three major rivers and many seasonal streams, the area is essentially semi-arid (*c.*500 mm per year (y^{-1})). The north-west tends to be wetter and the south-east drier, and this generates a seasonal gradient that drives the iconic animal migrations. The dry season runs approximately from June to October and the wet season lasts from November to May. Three habitats have been distinguished on the basis of this rainfall gradient. The short grass plains in the south-east are largely volcanic while the savanna areas dominated by acacia lie outside the volcanic influence. Savanna woodland, characterized by trees such as *Terminalia* and *Combretum*, have developed in the north-west with tall grasslands formed over infertile granitic soils.

The huge herds of ungulates (hooved herbivores) are the features that catch the eye. They can be roughly divided into migrants and residents. Of the migrants the wildebeest or gnu form the largest group with estimated numbers of around 1.5 million. There are also some 200,000 zebra and possibly half a million Thomson's gazelle making up well over two million animals selectively grazing different heights of grass (and thereby allowing many different species to occupy the same space). The resident animals comprise a very varied group from topi (related to wildebeest), antelopes such as kudu and impala, the graceful oryx, and larger animals such as rhino, giraffe, and buffalo (nowadays heavily poached). Predator–prey links are always in evidence—the protected status started in the 1930 in order to maintain a lion population but there are also leopards, cheetahs, and several smaller members of the cat family. There are numerous other predators and scavengers including hyenas, which are possibly the most abundant of the carnivores, wild dogs, and small animals such as mongoose and weasels. The animal life comes in a range of sizes from the largest land animal, the elephant, through aardvarks and hyraxes to elephant shrews. Overall the Serengeti ecosystem is renowned for its high biodiversity. This includes some 600 species of birds and a multitude of reptiles, amphibians, rodents, butterflies, and grasshoppers, although the immense variety of insect life is less well known. There is a remarkable array of beetles such as the dung beetle, whose efforts well exemplify the distributing and decomposing activities of the smaller organisms. Assessing such a mobile community of organisms, including microbes, to give a complete picture of the food web would be a monumental task. The simplified outline illustrated in Figure 22 indicates only the larger plants and animals.

The NCA is an extension of the Serengeti ecosystem. It became the first community-based conservation area in the world when it was formed in 1959 with a multiple land-use objective. It's unique setting in one of the world's greatest volcanic craters together with the vibrant wildlife and long-settled indigenous population (and site of one of the earliest human populations at Olduvai Gorge), ensures its place as one of the world's most important savanna reserves. The deep crater is about 20 km across and covers an area of around 300 km^2 with forested upper slopes and a valuable mixture of wet and dry vegetation (see Figure 25). The NCA contains around 25,000 large animals including the endangered black rhinoceros, numerous grazers, and their predators, with elephants and leopards at higher elevations in the forest. The annual wildebeest and zebra migration sees around 1.7 million ungulates move into the NCA in December with an outward migration in June. The area was set aside for the Masai

Savannas

25. Landscapes of the Ngorongoro Crater Conservation Area, Tanzania.

people, who number over 40,000, with the idea that they could sustain their way of life as pastoralists without impacting greatly on the natural rhythms of wildlife; however, as they have become more sedentary there have been increasing problems of population pressure.

Dry and wetland savannas. The Etosha National Park covers an area of around 23,000 km² in northern Namibia. Its name, meaning Great White Place, is a reference to the very large depression or pan which in the dry season evaporates and leaves pale coloured salts illuminating the savanna landscape. The main pan occupies an area of nearly 5,000 km² and is only filled with water after rains flood in from the north and east. The Fischer Pan in the east of the park holds water for all of the year. In these very dry savannas, such water holes provide a life-line for a very large range of animals that travel into the protected area from great distances around. There are also scattered springs bringing water from underground porous bedrock. The presence of water attracts animals of all descriptions from the very large (elephant, black and white rhinoceros, giraffe, and big cats), to middle-sized grazers and predators (notably zebra, many species of deer, warthogs, and spotted hyena), to numerous small mammals (e.g. mongoose, pangolin, porcupine) together with ground squirrels and reptiles (many snakes), and myriad insects. There are large numbers of birds ranging from perching birds and waders to vultures. The great attraction of Etosha is the seasonal theatre of animals, more widely scattered in the wet season when there are food resources over the whole of the park, and more concentrated in the dry season in a dazzling white dusty landscape around the water holes and springs (see Figure 26(a)).

Wet savannas are more like permanent wetlands, but the highly seasonal climate results in a rainy season waterscape and a dry season semi-arid savanna landscape. The Okanvango in Botswana illustrates this remarkable transformation from one part of the year to another. It is analogous to the Inner Niger delta in Mali, the Sudd region of the Upper Nile in South Sudan, or the largest of the world's wetlands, the Pantanal in South America (see Figure 26(b)). However, evaporation concentrates salts and the water changes in quality by the time the dry season arrives. It is estimated that about 36 per cent of the water is evaporated and 60 per cent transpired by the seasonally lush vegetation whilst the remainder percolates below ground. The water covers an area 250 by 150 km between March and June, and grasses flourish for a time with trees able to survive on slightly raised and better drained mounds—many being the result of termite activity. The wildlife is clearly dominated by the seasons but at its height there may be 200,000 large mammals with large herds of elephants and buffalo and antelopes (which have adapted to move rapidly through water and thus escape predators). The Okavango was formerly home for the indigenous Bushmen and Bantu, and the establishment of protected areas and the rights of local people have strongly influenced the concept of protection. The Okavango Delta is named after the delta-like features of the Okavango River. The river drains into the inland depression (an ancient lake in the

26(a) **Dry season savanna landscapes of the Etosha National Park, Namibia, showing the concentration of wildlife around water holes; (b) Dry season in the Pantanal, Brazil, showing grasslands that are engulfed by water during the wet season.**

Holocene) from the north, bringing a much needed seasonal relief to an area that would otherwise be more like the neighbouring Namib Desert. It is a very flat landscape with less than a 2 m variation in height. In some ways it resembles a giant oasis in the wet season, attracting wildlife from large distances around to the fresh water.

Savanna woodlands of Australia. The savannas in the north or 'top end' of Australia contain a number of national parks of remarkable diversity. Kakadu park lies some 170 km east of Darwin in the Alligators Rivers region of Northern Territory State and covers nearly 20,000 km^2. The world heritage listing is based on environmental and cultural grounds with a very rich history of aboriginal settlement, haunting cave art, and land management. To the east is Arnhem Land with further national parks to the west (Litchfield) and to the south (Nitmiluk). Although the vegetation includes monsoon forests and extensive wetlands, most of the vegetation consists of eucalypts and tall grasses supporting a wealth of plant and animal life. The main wet season or monsoon occurs between January and March, and the major hot dry period lasts from August to September. The onset of the wet season is often marked by spectacular storms whilst the intense dry periods of the year generate fires across the savanna. The indigenous aboriginal people identified six different climatic seasons, reflecting their deep understanding of the yearly rhythms and of their adaptations and management of the environment. Daily average temperatures vary between 20 and 350°C but rarely fall below 17°C. The frequency of fires has spurred many research projects into their nature and impact, including one of the world's largest fire experiments at Kapalga. The different fire treatments indicate surprisingly little change in the plant and wildlife composition so far, except for particularly sensitive locations around water courses or close to moist evergreen forest. This emphasizes the resilience of plants and much of the wildlife to burning, even at high intensities. However, as elsewhere in the savannas, fire exclusion results in increasing tree dominance, and many conservationists stress the need for areas to be protected from burning for longer periods of time.

The area covered by the national park is made up of several distinct landscapes although the lowlands and lowland hills covered by grassland and savanna woodlands occupy around 80 per cent. There are spectacular ancient volcanic hills and basins to the south, and a marked escarpment running for 500 km to the east delineating the 'stone country' of the serrated sandstone Arnhem Land plateau rising to 300 m with deep gorges. Some outliers of the plateau are the remnants of long erosion. There are also four major rivers and a vast network of creeks giving wide floodplains draining northwards to the sea, and tidal flats and mangrove at the coast that are internationally important wetlands.

Most of the Kakadu is covered by open, eucalypt-dominated woodland representing the last major expanse of relatively undisturbed savanna woodland left in Australia (see Figure 27). Abundant water during the wet season gives a green appearance to many of the trees and pockets of rainforest exist where they are protected from fire. The intense dry season can be extremely stressful and some plants such as the kapok bush (*Cochlospermum*

27. **Savanna landscapes in the Kakadu National Park, Northern Territories, Australia, showing the rocky outcrops and characteristic flat, grassy plains.**

fraseri) are deciduous and others semi-deciduous. The eucalypts, such as the 'Darwin woolleybutt' *(E. miniata)* that can grow to 20 m, survive well because their deep rooting systems can tap underground moisture. The understory is usually made up of smaller trees such as the widespread pandanus *(Pandanus spiralis)*, the green plum *(Buchanania obovata)*, and distinctive plants grasses such as spear grass *(Sorghum* species) that can reach up to 4 m in height. There is a rich diversity of plants in the ground layer with many types of sedge and wildflowers as well as numerous grass species. All told, there may be well over 1,000 species of plants and some estimates suggest around 2,000.

With such a variety of habitats within and surrounding the savanna, it is hardly surprising that there is a rich diversity of animal life and it is strongly influenced by the seasons since much of the lowland is flooded in the wet. Although there are many links with South-east Asia, the fauna has evolved largely in isolation and many of the animals are unique to Australia, especially the great variety of marsupials. Small mammals are numerous as well as more visually obvious larger animals such as kangaroo, wallabies, spiny anteaters, and dingoes. Flying foxes, possums, and rodents are more active at night, which also sees the activity of numerous bat species. Birdlife is particularly impressive with perhaps one-third of Australia's total (around 280 species), and much is concentrated around watercourses. Amongst the most visible of the birds are parrots, honeyeaters, lorikeets, and cockatoos. Insect life is very profuse, and estimates indicate that there are over 10,000 species, ranging from noisy cicadas to a multitude of butterflies, moths, flies, and beetles—many of which participate in the vital processes of pollination and decomposition of organic matter. There is also a wide variety of reptiles including skinks, pythons, and lizards. There are no large predators as found in the African savannas; one of the largest being the dingo, which probably came from east Asia a few thousand years ago. Termites are major components of the landscape by harvesting and decomposing organic matter, thus redistributing nutrients throughout the landscape, and helping to store moisture. In doing so they promote greater biodiversity and can even be said to restore and renew their environment. At the same time because of their widespread wood-consuming activities they lead to them being regarded as pests where they are found close to human settlements.

Chapter 5
Savannas and human evolution

Savannas have helped to shape the evolutionary pageant of human history and the dispersal of our ancestors across the continents. Although there is considerable controversy over the causes and mechanisms of this evolutionary story, ecological, environmental, and genetic evidence suggests that human-like primates first arose in savanna sites within Africa. These habitats offered more open conditions of grassland and woodland that are believed to have stimulated the development of specific physical attributes and social behaviour leading to advanced and mobile societies. The early history of scavenger-hunter-gatherers and their migrations out of Africa into Asia, Australasia, and eventually into the Americas makes a compelling saga. Modern humans contended with climatic oscillations and changing, constantly challenging environments in their migratory spread over the Earth's surface. The wide panoramic landscapes of savannas offered resources and opportunities that facilitated these extraordinary movements.

The origins of savannas

Savanna vegetation is characterized by a ground layer of grasses and associated herbaceous plants that forms a defining component of the landscape. Although the emergence of flowering plants (*angiosperms*) can be dated back to the early Cretaceous around 140 Mya, the grasses then formed only a small proportion of the total flora. The savanna landscapes as we define them today can be traced back to the evolution of grasses in the late Tertiary some sixty to seventy Mya. The large grass family (Poaceae, with around 11,000 species) became divided into those that flourished in more temperate climates (C3) and those that evolved to deal with warm climates (C4). It is likely that the grasses originated in the old Gondwana land masses as C3 shade-adapted plants but rapidly expanded from the Eocene (fifty-five Mya), spreading into open habitats by forty Mya. This was probably not a simultaneous event across the globe, but was first evident in South America (Eocene) and lastly in the Australian savannas (late Miocene-Pliocene).

Grasses with C4 metabolic pathways dominate tropical savannas today and account for possibly 30 per cent of global C-fixation. The physiological mechanism, referred to as a CO_2-concentrating pump, enables the plants to have a high photosynthetic efficiency in warm climates where there are low natural CO_2 levels. The C4 pathway proved to be a major evolutionary success, giving rise to grass plants that have been crucial to human development. These span a huge range of economically and culturally important plants such as maize, *Zea mays* first cultivated in Mesoamerica around 2,500 years ago, sugar cane (several species of the genus *Saccharum*), sorghum (from the sub-family Panicoideae), or the largest and fastest growing member of the grass family, bamboo (sub-family Bamusoideae) with around ten genera and nearly 1,500 species through the world. The same mechanisms have been identified in other, mostly herbaceous, plants such as sedges (Cyperaceae) and numerous other well-known plant families such as the daisy (Asteraceae). It has been widely accepted that the C4 pathways evolved as a response to declining levels of atmospheric CO_2, making a CO_2-concentrating mechanism a valuable adaptation to increase photosynthesis. However, this simplified explanation has been criticized and it appears that a combination of factors aided the rise of the C4 grasses. These included the influence of seasonal climates with strong dry episodes that led to water shortages and facilitated fires, especially in monsoonal regimes marked by a high incidence of lightning strikes generating fires at the stormy onset of the wet season. The influence of the increasing number of herbivores could also have played a part, since browsers would have reduced the leaf canopy of woody vegetation and limited re-growth. It is not clear whether the process of change over from C3 to C4 was a single evolutionary event or whether there were multiple mutations and possibly reversals as climate varied.

The rise of C4 plants was rapid in terms of geological timescales, starting in the late Miocene (five to eight Mya) and extending into the Pliocene (two to eight Mya). The same processes have continued since that time through the constant variations in climate. Direct plant fossil remains have long since decomposed and evidence for these changes is largely derived from C-isotopic data in organic residues and fossil soils. The transformation was accomplished rapidly across all the continents and the proportion of C4 grasses is believed to have risen from around 0 to 80 per cent in two to four million years. Since the takeover occurred globally at around the same time, it seems likely that there was a global reason, such as climate and levels of atmospheric CO_2 concentration. The early grasses adapted to seasonality and warmth by developing drought-resisting features. The main growth points (*meristems*) are located at the ground surface or even within the soil, with increased root and decreased shoot growth, sunken stomata, and thick cuticles which help to resist fire. There is also strong evidence to indicate that the rise of grasses coincided with the spread of mammals since the grazers and browsers became adapted to a diet high in cellulose and silica-rich leaves and in this sense savannas represent an example of co-evolution. Other dryland plants also developed in response to seasonality such as drought-resisting shrubs. The typical savanna vegetation with high proportions of grasses and shrubs became evident by the Miocene around sixteen to eleven Mya.

Origins and movements of people

While there has been much debate about the part played by savannas in influencing the course of human evolution, the major outlines have broad agreement. The greatest numbers of the earliest hominins (belonging to the family Hominidae, with a single genus *Homo*) are to be found in sub-Saharan Africa, and it is generally agreed that the closest relations of modern humans evolved over large parts of that continent some two to five Mya. Evidence of these early ancestors has been found in eastern and southern Africa and also in areas as far apart as Malawi and Chad. The best known fossil evidence of the earliest remains has been discovered in Kenya, Tanzania, Ethiopia, and more recently in South Africa. The largest single fossil hominin assemblage found so far has been termed *Homo naledi* in the Rising Star cave system 50 km south of Johannesburg. It is thought that the open landscapes, but especially the more wooded savannas, offered opportunities for early pioneers to become hunter-gatherers and scavengers, and to range over extensive tracts of land. This is likely to have been the result of a combination of propitious environment (food resources and water) together with the development of advanced forms of social behaviour (integrated activities and the use of tools). It is believed that the first members of the genus *Homo* appeared 2.5 Mya. At around two Mya there appear to have been six human-like species, although the evidence is scattered, fragmentary, and open to constant re-interpretation. It was clearly a prolonged and complex process of evolution and while the mosaic of savanna environments provided a context, it is likely that the concentrations of early hominins were connected with the most favourable habitats having a proximity to water such as forest edges, wooded patches, and gallery forests.

From fossil evidence and information derived from a wide variety of methods such as isotopic dating, palaeomagnetism, amino acid analysis of organic materials, and genetic tracking, it would seem that the earliest primates and hominins ranged over a mosaic of environments. One authority commented that there is no evidence, in the early stages of human evolution, for an environmental change that would determine a spread into savanna. Nevertheless, some sites, such as the Olduvai Gorge in northern Tanzania, have been shown to reveal recurrent ecosystem variations. Open C4 grasslands changed relatively abruptly (over several hundreds to thousands of years) to closed C3 forests with periodic reversals. It does appear that there were climatic variations with markedly drier and wetter episodes in at least some of the areas of early hominoid activity, and this may have helped to trigger development.

It has been suggested that biological and human evolution, which proceeded in fits and starts, was at least partly occasioned by abrupt climatic changes stimulating or forcing new adaptive strategies. The open landscapes would have provided a broad spectrum of food resources whose exploitation required and has been correlated with increases in brain capacity. There is evidence of an increasing adjustment to the ground level through the development of bipedalism. This freed the use of hands and provided greater adaptability to the demands of seasonal rhythms. Over recent years there has been an emphasis in the literature on the growth and increasing sophistication of social systems, including the increasing power of communication through language. Savannas provided a range of environments that are likely to have stimulated such development. The evolution of language would have required an ability to communicate beyond a level possible with very small groups, and would in turn have accentuated social customs. Although some of more advanced human-like individuals such as *Australopithecus afarensis* seem to have been at home on the ground and in trees (for instance, the fossils found in the Hader area of Ethiopia), there are clear signs of later, greater adaptations to more open and savanna-like conditions. The detailed description of early hominins within the Olduvai Gorge, where there is a two-million-year sequence of lake and river sediments, illustrates this evolutionary panoply. *H. habilis* appeared in the Olduvai area around 1.9 to 2 Mya, with *H. erectus*—from 1.2 Mya—and *H. sapiens* very much later (see Figure 28). Broad-scale climatic changes continuously varied the landscape over these timescales. Some regions became wetter, such as the southern Sahara, as evidenced in the ancient stream channels now buried by sand. Equally there were periods that were markedly drier with extreme droughts when there would have been inadequate food resources.

The explosion of information in evolutionary genetics over the past few decades has offered further evidence on these origins and migrations. The genetic record provides a life story of living organisms, where the genomes of living individuals are passed down from ancestors and where ancient DNA is preserved in organic remains. It is claimed that it is now possible to catalogue the genetic differences between humans and our earliest relatives comprehensively, and to trace one layer of history on top of another. This evidence has confirmed that the first hominins evolved in Africa from six to seven Mya and that bipedalism had developed by between four and six Mya. The first humans appeared around 1.9 Mya and the speed of migration was then extraordinary since *H. erectus* has been recorded in East Asia by 1.8 Mya. The use of fire has been dated back around a million years in Africa. Anatomically modern humans appear over the last 200,000 years and differ from ancestors in their morphological structure and brain, in their social organization, and probably in language as well as in selected genetic changes. There is also evidence of a high level of genetic differentiation over time between relatively close populations. This has been interpreted as being the result of obstacles to dispersal such as mountain ranges or powerful water barriers. On the other hand the open savannas and coastal plains offered greater visibility, food and water resources, and accessible routeways. Integrating geographic information such as this with the genetic information will provide a powerful approach to interpreting genetic diversity in the future.

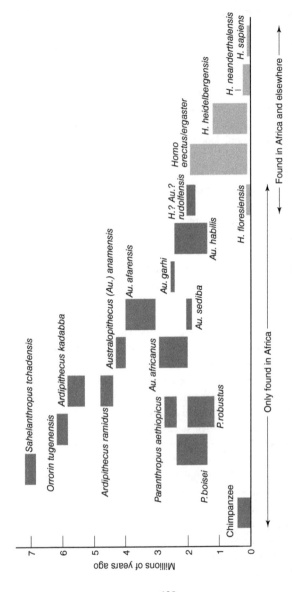

28. The human ancestral tree and evolution from fossil hominins. Recent discoveries at Rainbow Cave 50 km south of Johannesburg have been named as *Homo naledi* and may represent some of the earliest of human remains.

Dispersal and migrations

The dispersal of early humans away from their savanna-like origins in Africa makes up an epic narrative and took place over a remarkably short period of history—around 100,000 years. Current thinking indicates that some African fauna moved out in the warm climates of the last Interglacial around 110–130 Kya and that early hominins expanded about the same time (ninety to 130 Kya). A likely route out of the African continent would have been the region around the Awash Valley in Ethiopia, which is adjacent to the southern end of the Red Sea. This passage would have afforded a short crossing to the Arabian Peninsula and then along the coast to the Indian sub-continent. An alternative route further north might have taken early migrants into the Middle East and then east via the Caucasus, Black, and Aral Seas, and on into eastern China. The sea crossings would have been easier to pass at times when glaciations in the northern hemisphere had resulted in lowered sea levels. At different climatic phases, the migration routes would have been facilitated when increased rainfall promoted savanna-like vegetation in what are today semi-arid regions. Equally there is evidence of periodic dry episodes that would have halted expansion, cutting off routes with desert-like conditions.

There are differences of opinion as to whether there was one major route or multiple routes, and whether there was a continuous movement or pulses of migration in favourable periods. Whatever the route and whatever the number of movements, the savanna landscapes provided a range of favourable habitats for expansion. The migrations appear to have reached South-east Asia relatively rapidly and this region is claimed to be only second to Africa in the length of recorded settlement. Although it is not proven, it is suggested that the first hominins to reach South-east Asia were ancestors of *H. floresiensis* with similar physical proportions to those of *H. habilis* in the Olduvai Gorge. *H. erectus* appears in the record in Java and a few other sites only 100,000 years after their

record in Africa. Anatomically modern humans are identified in
the region from sixty to 50,000 years ago and current genetic
evidence for *H. sapiens* tends to favour single migration out of
Africa some fifty to seventy Kya.

The migratory movements passed through phases of expansion
and bottlenecks that were largely driven or limited by climatic
change, and shaped by the configuration of the land masses. By
the time early migrants reached south and south-eastern Asia,
many of the landscapes were drier than at present notably during
the maxima of Pleistocene glaciations. Present-day Sri Lanka was
attached to the Indian mainland and much of South-east Asia was
joined into one land mass (Sundaland), separated from Australia
and New Guinea (Sahul) by deep channels. These breaks in the
landmass with deep, fast flowing currents, known as Wallace's or
Weber's lines after the pioneer naturalists, severely restricted
human and animal movements. Sea levels were estimated to be
over 100 m lower than the current distribution, exposing land that
is today under the sea. The extent of open savanna-like conditions
was almost certainly much greater than the present day although
it is probable that the main migration routes to the south-east
were along coastlines and may have involved some of the earliest
known seacraft (see Figure 29). The theory is that the expanses of
open vegetation in Africa and elsewhere supported a proliferation
of food resources including large herbivores and many smaller
animals. This is believed to have stimulated the development of
tools to catch and prepare food. Changes in the vegetation as
shown from pollen data, indicate a transition from pine-alder
associations at glacial maxima to broad grassland corridors with
gallery forests along water courses as the climate became drier
and more seasonal. This in turn attracted characteristic savanna
animals such as elephants, rhinoceros, and deer.

The area known as Sundaland in South-east Asia, exposed by the
falling sea levels, was nearly twice the area of the present day land.
Many of the islands making up present day Indonesia were joined

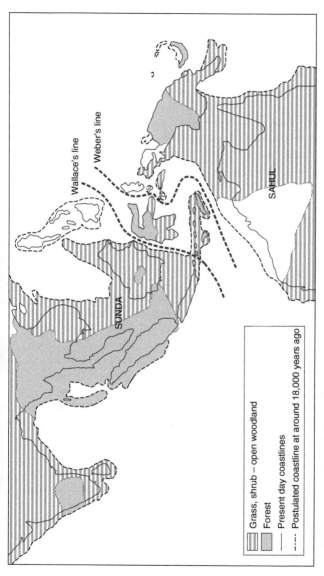

Grass, shrub – open woodland
Forest
Present day coastlines
Postulated coastline at around 18,000 years ago

29. **Habitats crossed by early migrants in south and east Asia at low sea levels during the Late Glacial period, illustrating the extensive stretches of exposed land.**

together facilitating movement of early humans. Oscillations of sea level, the most recent of which was a rise around fourteen Kya, eventually cut off these pioneer groups which are likely to have been very small and they subsequently evolved in relative isolation. Many of the signs of these first venturers would be located below present-day sea level. Over higher land it appears that sites have been buried by the active volcanic debris typical of the island arcs of South-east Asia (for instance the catastrophic eruption of Toba volcano in Sumatra at around seventy-four Kya). The coastal route from Africa to South-east Asia and then on via the island arcs to Australia is probably the principal migration passageway. However, it is likely that there were multiple expansionary movements of peoples taking advantage of the open landscapes of woodland-scrub-grassland that existed over large parts of Asia in the moister climatic phases.

Passage to Australia

Although the crossing from the nearest South-east Asian islands appears straightforward with relatively short distances between them, the straits are frequently deep and treacherous. There is little doubt that early migrants did cross significant stretches of sea but they were believed to be mostly small groups and often became isolated. Despite this, some genetic evidence proposes that the first group to reach present day Australia was a substantial number (possibly around 1,000), which would represent a surprisingly large migration. Movements of this size would contradict the idea of very small, gradual, incremental, and episodic shifts of people. The closest part of Australia to places such as New Guinea would only have been a little over the line of sight at the times of the lowest sea levels in the past. It has been suggested that there might have been indications of land, such as the appearance of land-based birds and a dust haze suggesting a landmass. There is evidence to show movement between the islands (and later out into the Pacific), so the early colonizers might have acquired the ability to cross by sea. There is no suggestion that *H. erectus* ever reached as far as Australia, and there are several different dates put forward for the first landfalls although they are likely to have been between forty-five and seventy Kya. The first settlers in present day Australia would seem to have been in Arnhem Land, where artefacts have been found about 50 km inland dating to sixty-one Kya ± 10. However, the initial landing sites are almost certainly under the sea today and were maybe 200 km further north of the present coastline. There were numerous food resources including a number of very large animals (such as the now extinct giant flightless birds and kangaroos). These are vividly depicted in the cave art for which the indigenous areas are famous.

The early migrants have been described as supernomads. They were able to cover extremely large areas remarkably rapidly. They appear to have adapted to the varied landscapes, particularly the more open savannas that resulted from the oscillations of climate in the post-glacial period. For instance there are palaeo-records showing widespread flooding from eighteen Kya, with shrub vegetation seemingly predominant at around thirteen Kya and forest areas becoming dominant from ten Kya. The present day monsoonal regime of northern Australia appears to have developed by fourteen Kya and extensive savanna (woodland and grassland) by around eleven Kya. Most of the coasts, open landscapes, and forest were occupied to some degree by thirty to forty Kya. Adaptations to the semi-arid landscapes took longer but seasonal rainfall and more moist periods in the past appear to have allowed access and even settlement in favourable locations. Essentially these colonizers were highly mobile and adaptable to the changing climatic regimes. The savannas seem to have acted as a conduit for the migrants who utilized the mosaic of forest, grassland, and desert—as long as there was water, food, and wood for fuel. Savannas seem to have been burnt from the earliest times (e.g. by travelling groups carrying the traditional firesticks), and this may have exacerbated the increasing dryness. For example, Lake Eyre shows a record of a tree-shrub-grass mosaic being transformed into a desert landscape some sixty to forty-five Kya.

Passage to the Americas

Unlike the model described for the Old World, savannas played very little part in the settlement of peoples across the continent, although they may have played a part in promoting movement. The widely accepted theory is that the initial colonization of the Americas took place via the Bering Strait from Asia at times when the last glaciations had locked up much of the oceanic water and sea level drop meant that a wide causeway (Beringia) existed from what is now eastern Siberia to Alaska. Although there are opinions that some migration was earlier, and some have suggested that it was partly trans-oceanic, most authorities agree that the dispersal into what is now North America took place after around seventeen Kya. The movement of small groups of people out of Africa, through the Middle East, then on to the north-east of Asia, and eventually into Beringia was inevitably a slow and tenuous affair slowed by distance, climatic changes, and environmental barriers. The onward movements spread over the continent may have reached as far south as Mexico, but as far as the South American continent is concerned the main direction would have been a southerly route via the western coast or along the flanks of the Rockies.

Early migrants passed down through Central America, probably following the more accessible routes via higher land or along the coast. There is only very scattered evidence of these first groups, and the earliest dated remains (rock shelters, cooking hearths, maize cobs) are arguably from nine Kya. At lower altitudes where the climate was more typically tropical, there is a nearly continuous series of savanna patches on elevated plateaus, and less commonly over lowland plains closer to sea level. These appear to have been utilized and settled at later stages in the migrations. There is evidence of severe droughts and animal extinctions similar to the movements through Africa and Asia, and the impact of climatic changes appears to have caused analogous bottlenecks to migration. The pioneer colonizers of the tropical parts of Latin America appear not to have occupied the savannas as much as the forest environments or temperate uplands. The great civilizations that grew up in the continent appeared at a later date. They tended to be located on higher ground such as the Mexican plateau (Aztec and their predecessors), and followed the line of the Rockies-Andes cordillera (Inca and pre-Inca groups on the Pacific coastal plains). These early groups of peoples moved very rapidly over the higher land to the south of the continent. Other groups settled in the humid Atlantic coastal plains of Central America (Olmec and Maya) and, perhaps by following the coastline, in Amazonia. There are scattered savanna tracts throughout the rainforest with seasonally flooded formations along river courses (*várzeas*), on Marajó Island and coastal areas in the mouth of the Amazon. The densest occupation by indigenous groups appears to have been in the eastern forests, particularly alongside rivers. The early settlers over the savannas would have needed access to water resources. Recent research has shown distinctive genetic episodes for these initial movements of people and it is believed that some of the pioneer Amerindian groups became isolated quite early in the migrations.

Although the tropical savannas were probably not the principal routeways in South America, there is evidence that some of the colonizing groups found ingenious ways of utilizing more favourable sites where there was water. There is evidence of extensive occupation with drained and raised fields in western moist savannas such as the Llanos de Mojos similar to the land uses in a number of lowland Mayan sites. Around the Amazonian fringes, Amerindian groups like the Kayapó, appear have deliberated transferred soil over short distances from forest to savanna in order to plant different species and diversify the sites. The open landscape was probably used more for access and hunting. Nevertheless, the Central American, *llanos*, and *cerrado* landscapes were not settled by substantial groups until much later than the African model and remained relatively sparsely populated until colonial times and later.

There have been remarkable stories of migration in almost every historical era and it has been suggested that all the major language groups were in existence by around 30,000 years BP. The savanna biome has evolved relatively recently in geological history and the rise of humans has only occupied a very short fragment of this time. As climate has varied so the forest-savanna and savanna-desert boundaries have shifted to and fro with oscillations of rainfall and temperature. The arrival of the first humans accentuated the impact of fire, constantly gnawing at the forest frontier and renewing the grass-dominated plant life. Forms of agriculture appeared around fifteen to five Kya and developed independently in several regions following the end of the last glacial period in the northern hemisphere. The area from Egypt to western Asia through to India probably saw some of the earliest sedentary forms of agro-pastoralism, often based on primitive grasses. The Fertile Crescent in the Middle East is credited with the development of wheat, the mid-Nile with sorghum, while South-east Asia witnessed the cultivation of rice. Pastoralism also has an early history with domestication of cattle and possibly the gradual incorporation of wild animals into a more domestic way of life.

Chapter 6

Changing patterns
in the landscape

The relentless growth of population throughout the world and particularly over the past century has significantly modified savanna landscapes, though the proportions and directions differ from continent to continent.

The appearance of many savannas has been greatly affected by evolving ways of using the land. Many of the landscapes in the New World have only been occupied for relatively short lengths of time and settled only in the most favourable locations. The Old World by contrast, from Africa through India and east Asia to Australia, has experienced a long history of nomadic movement and occupation. The generally inhospitable nature of the soils and the variability of the climate have not been conducive to permanent settlement. Hunting wildlife, foraging, and collecting from the wide variety of plant life constituted the main land use for thousands of years. This pattern changed with the arrival of colonists from Europe and land tenure as well as landscape appearance changed from largely locally self-sufficient communal types of land management to one of large estates and commercial scales of development. The discovery of minerals and the development of transport networks have opened up great swathes of land that were isolated for thousands of years. With the advent of modern farming techniques and very different ways of managing woodlands, together with larger scales of operation and greater capital investment, savannas have been drawn into a global interchange of goods and services.

Landscape change in Latin America

The Brazilian savannas. Before the arrival of the Portuguese in 1500, the *cerrado* was sparsely inhabited. The indigenous people made use of whatever plant resources were available and hunted the wide range of animals whilst keeping within range of water supplies. The sites close to rivers, moist depressions, and gallery forest were able to support small-scale agriculture but the seasonally parched plateaus were unsuitable resources for farming. Consequently large stretches of central and western Brazil sustained relatively few people. Colonization reduced the numbers of original inhabitants even further, and landscapes became dominated by very large estates where the principal land use was low-density cattle-rearing. The waves of expansion of population in Brazil took place from the initial coastal settlements towards the interior. In the early days mining played a major role in extending development into the *cerrado*, such as the iron ore and gold industries in the State of Minas Gerais (literally 'general mines'). Much of the interior was hardly developed at all (known as the *sertão* and similar to the *Outback* of Australia) until the creation of the new capital, Brasília, in 1960.

The gradual transfer of power from the former capital in Rio and the spurt in policies emanating from the Federal District has attracted immigrants from the poorer parts of the country, particularly from the drought stricken north-east. At the same time, the irregular water supply over the *cerrado* and the adverse soil properties limited agricultural development. This persisted until the 1970s when the problems of low nutrient reserves, fixation of available phosphorus by iron oxides, soil acidity, and aluminium toxicity were overcome through liming, fertilization, and irrigation. This transformed the woody grasslands into vast fields of crops such as cotton, maize, dryland rice, beans, cotton, and soya. Parallel improvements in livestock management encouraged extensive herds of cattle and Brazil is now reputed to have the greatest cattle numbers in the world, although this has spread from the savannas into the neighbouring forests (see Figure 30).

Large commercial developments required major capital investment, and this was promoted vigorously by the government, with a strong emphasis on the development of export markets. As a result, land ownership has been further concentrated in the hands of a small proportion of owners. Recent estimates indicate that something like 75 per cent of the land is owned by around 10 per cent of the population and the agricultural frontier has steadily moved west and north. Small-scale and subsistence farming remain at a disadvantage because of the costs of fertilisation, pest control, and difficulties of gaining agricultural credit. Despite this the smallholders have shown a remarkable resilience and in some areas horticulture has flourished. This is notable in the river valleys and better-watered areas close to urban centres (e.g. the *cidades orientales* near Brasília—so called because of the large proportion of farmers of Japanese origin). There are still limitations of fuel supply for domestic and industrial use, and for many years the sight of trucks carrying charcoal to fuel the steel industry was a common feature of the rural road network. This also expanded remarkably with the development of the central plateau and the roads linking the Trans-Amazonian network and the western states such as Mato Grosso and Mato Grosso do Sul. The energy deficiency also triggered a huge increase in hydroelectric schemes, with estimates suggesting that 90 per cent of Brazilians depend upon hydroelectric power for their energy. The retaining dams pounded-up water for storage reservoirs and have also been used to develop a freshwater fishing industry.

The landscapes have been transformed from virtually undisturbed savanna to a sea of industrial-scale agriculture over a period

30. Extensive cattle ranching over the Brazilian *cerrado*, showing the nearly complete conversion from the original wooded savanna.

of a few decades. The *cerrado* has frequently been described as the world's last great agricultural frontier; so the process of converting the savanna landscape and the increasing incidence of anthropogenic fires raises a number of concerns. Conservationists wish to conserve a greater proportion of the natural vegetation and wildlife as less than 3 per cent is under some form of strict protection at present. Yet the startling rapidity of development and occupation of the *cerrado* has led to an inevitable increase in the pace of savanna clearance and augmented degradation of limited soil resources. There have been increasing struggles over land and rising violence. Over recent years there has been an increasing 'landless' movement similar to that in the Amazon region, and the MST (*Movimento Sem Terra*) or movement for those without land, is now depicted as the largest social movement in Latin America. In addition to their demand for land, the MST also advocates agro-ecology, as opposed to blanket commercial farming, and greater food security and control. There are also concerns over the provision of water supplies, for domestic and commercial use and to satisfy the growing demand from agriculture. Groundwater levels have dropped in a number of localities and long dry seasons provoke demand that is difficult to match. Equally demanding of land is the politically inspired drive to occupy and develop the western borders of the *cerrado*, partly to consolidate the territory in areas which have seen international disputes in the past. At the same time there has been a significant growth in the influence of NGOs in all aspects of life and this has gone along in parallel with the democratic changes in Brazil since the military rule ended in the 1970s, and the emergence of left-leaning governments.

The major concern is the rate of savanna destruction and an estimated 66 per cent has already been transformed. Since the rainforests (both Atlantic and Amazonian) get most of the publicity both at home and abroad, savanna conservation has been less prominent. Despite widespread knowledge that the *cerrado* has the richest of all savanna vegetation the progressive loss of land to agriculture has not halted, while construction and urbanization have increased dramatically. Savanna conversion has impacted on wildlife, with World Wildlife Fund (WWF) estimates of sixty native animal species now vulnerable, twenty endangered, and twelve critically endangered. The giant anteater and maned wolf are examples of animals that are on the critical list; the anteater is particularly vulnerable since it has a low reproduction rate and limited diet, and is affected by the increasing incidence of fires. Habitat destruction is having a major adverse effect on land animals with the possible exception of the diverse and more mobile bird species.

The llanos. Although the savannas of the *llanos* and *cerrado* are similar in many respects, the landscapes of the Orinoco Basin are dominated by grasslands. Despite the fact that the savannas of Venezuela and Colombia supported significant populations of Amerindians, particularly along water courses, their numbers plummeted after colonization and subsistence ways of life have remained at the bottom of the social hierarchy. The introduction of cattle farming by the Spanish after the mid-16th century remains by far the predominant land use.

As in the *cerrado,* land tenure was characterized by large estates and this extensive type of management has continued today despite a number of attempts at land reform. The poor nutrient value of much of the savanna has limited stocking rates and productivity, although the very varied nature of the environment means that there are favourable and less promising areas for agriculture. Forage production is low and technical input has been lacking in comparison with the developments in Brazil over the past few decades. Cultivation is limited with traditional crops of maize and beans with low yields and attempts to diversify have been constrained by a lack of capital input. Commercial agriculture has increased with crops like oil palm and rice, and a problem arising from this intensification has been the increase in agricultural toxins in the runoff. One of the more unusual forms of resource exploitation of the wetter more swampy areas has been the management of the capybara (see Figure 18). The capybara can develop large colonies of around 200 individuals/km^2 and, although a rodent, can be managed like cattle as a ruminant.

One of the problems in land close to low-lying water courses is the seasonal flooding. This is followed by water deficiencies in the dry season, and attempts are being made to channel the excess water in the wet and conserve it for the dry season. The development of water control systems with dykes and waterways has led to the drainage of much of the wet savanna. This has threatened the rich wildlife in these areas although it has provided water resources for agriculture during the dry periods of the year. Clearance of the savanna woodlands is a serious issue and afforestation using exotics like the Caribbean pine (*Pinus caribaea*) and Eucalyptus species has contributed to the decline in the native savanna fauna. Wildlife in general is given little protection and only a small proportion of the *llanos* is protected (*c.*4 per cent in Venezuela), with the same problems of monitoring and policing as is found in the *cerrado*. The discovery and exploitation of oil, affecting nearly 3 million hectares in Venezuela, seems to have had little beneficial effect on the way of life for the majority of the rural inhabitants and has knock-on effects on the environment through habitat fragmentation, road and settlement construction, and air/water contamination.

Landscape change in Africa

The African continent enjoys a vast array of differing savanna landscapes, from semi-arid margins south of the Sahara and around the Kalahari deserts, to moist woodland savannas abutting on to evergreen forest and spectacular wetlands. The varied environments and convoluted historical patterns of land use have produced widely different arenas of human activity.

West Africa. These distinctive regions have been utilized for thousands of years, mainly for hunting and gathering, then evolving into agricultural and pastoralist communities with conflicts between the two ways of life. The marked latitudinal changes in climate have also meant that more southerly areas with more favourable rainfall maintained cultivated crops more successfully than the Sahel regions to the north.

The fragmentation of land holdings and persistence of old land tenure systems run by communal groups has resulted in much of the small-scale farming existing at a subsistence level. The staple crops are sorghum and millet in the drier north while maize and rice are more prevalent in the south. At the same time there have been government-backed moves to develop large-scale commercial farming for export crops such as cotton or groundnuts. Livestock rearing is particularly important in the Sahel region such as northern Nigeria and Cameroon, Mali, and Burkina Faso. A further issue is the lack of energy sources particularly in the north where fuelwood has been the traditional source. Planting of exotics, notably species of eucalypt, has inevitably impacted upon the local biodiversity. One of the problems for wildlife conservation is the extent of hunting which has not only reduced the populations of some vulnerable species like the black rhinoceros and elephant but also resulted in the extinction of some small grazers. Although the region has a number of well-established national parks and protected areas, the level of protection is often nominal.

Competition for land is a rapidly growing problem as there has been serious loss of wooded resources, considered to be a critical aspect of the decline in livelihood resources. A study across the whole of West Africa for a decade from remote sensing, showed a loss of about 1 per cent annually in dense woodland and around 0.5 per cent for more open wooded savanna. A study in Ghana over ten years to 2001 showed an 18 per cent loss of woody savanna and shrubs converted to an herbaceous cover, although 'farmed parkland' acts as some measure of conservation. A related decline is evident in the availability of non-timber forest products for fuel, fruit, and a host of other uses. The most important local tree species have been identified as *Parkia biglobosa*, *Vitellaria paradoxa*, and the spectacular baobab (*Adansonia digitata*). Baobab trees make up genus with nine species ranging from West Africa to Arabia and Madagascar (and one native to Australia). The baobab has been shown to be well preserved in traditional communal areas due to its longevity and low mortality rate. The fruits produce seeds that can be dried and powdered for trade. Its striking structure (the so-called upside down tree) and water-holding capacity (as much as 100,000 litres) provide an enduring image of the drier savannas (see Figure 31). They are bat-pollinated and consequently the disturbance to the bat habitats is an issue of growing concern. Shrub-dominated areas appear to be more tolerant of land use changes than tree savannas.

Recent work in Mali demonstrates that all the potentially useable land has now been exploited. An idea of the vulnerability of the farming system in these drier regions is evidence showing that farmers may expect a period lasting from one to three years when fertility is too low for cultivation, in addition to the concern over the erratic nature of the rains. Where local people have control over their land resources, they clearly try to manage their resources more effectively to their own advantage, but these local controls are overshadowed by large-scale commercial enterprises. A related consideration is the impact of reducing the vegetation cover on the ecosystem balance, including the reduction in organic matter which acts as a nutrient and water reservoir. Clearance of vegetation has increased runoff representing both a loss of water resources and a periodic damaging effect of water discharge. Clearance of savanna in Niger and Lake Chad catchments for example, was reported to have increased streamflow by between 33 and 90 per cent. Overriding many of these concerns has been the discovery and exploitation of major mineral and oil resources. The highly productive oil industry in

31. The Baobab tree landscape of Madagascar, or the so called 'upside down tree'.

Nigeria for example has not seen a channelling of resources back to the less affluent north, and this has accentuated the ethnic differences both in Nigeria and a number of other countries in the region. The major concerns at the present time are political instability, incidence of disease such as Ebola across much of the humid savanna, lack of fuel for cooking and reliable water supplies, drought plus desertification, and land degradation in the north.

East Africa. Hunting, gathering, and foraging ways of life are still relevant in many parts of Africa, although agriculture has been a feature of the landscape for 2,000 years or more. Semi-nomadic pastoralism has also been an important feature with an up-and-down history as a result of diseases such as rinderpest (now hopefully extinguished) and trypanosomiasis. Periodic droughts have also a long history of afflicting the livestock herds in dry savannas. Colonization changed the landscape by bringing in settled townships with an extensive infrastructure, and network of roads and railways. The older forms of land tenure, typically community-based, were replaced in many areas by large-scale commercial farms. Different countries have reacted differently following independence and Tanzania for example, developed a socialistic land system following independence in 1964.

The land uses vary across the region although the majority of the population is still based on small-scale farm units. These produce staples like wheat, sorghum, and millet depending on the climate and soils with many subsistence vegetable crops. The main export crops are coffee and tea. Perhaps the most striking feature of the savannas in the region is the emphasis on ecotourism, based on the prolific wildlife outlined earlier and strengthened by the varied number of protected areas. The side effects of land management are, however, sometimes unexpected. For example the removal of large grazing herds has been shown to lead to an increase in small mammals followed by a less welcome increase in predator snakes.

Some of the major concerns at the present time are the clashes between land used for wildlife and the growing human population, the availability of energy sources, climatic variability leading to water deficits and droughts, poaching, and the widespread problem of land degradation.

Miombo savanna woodlands. Land covered by *miombo* vegetation has been utilized in one form or another since the earliest incursions of humans. The characteristic trees such as *Brachystegia* and *Jubernardia* were used for construction and for fuel, and are valuable for their coppicing regeneration capability. Shifting agriculture was widespread along with hunting and gathering. The pioneer forms of land management were characterized by communal properties led by the elders of settlements, a form of authority which persists today in many areas but with varying degrees of pressure and success.

Much of the *miombo* has been affected by human activity and it is probable that very little of the original vegetation remains. Despite the traditional land use in the form of small-scale agriculture, a great variety of plants has been cultivated ranging from cereals such as millet and sorghum, together with cassava, groundnuts, gourds, yams, sweet potatoes, and numerous vegetables. The *miombo* soils are inherently more fertile than their South American equivalents, being less leached and developing from a range of parent materials. Consequently there has been a greater proportion of commercial agriculture, for instance tobacco and maize, in those locations (such as the inclusions of *vley* or *dambos*), where the moisture regime as well as the nutrient status can be managed effectively. Human population densities and the numbers of livestock in the *miombo* region are low in comparison with those in other African savannas. This is partly a result of generally infertile soils, vegetation with little nutritional value, and disease (such as tsetse fly). Despite these restraints the *miombo* resources are important for millions of people both living in rural areas and in urban settlements. The parts of the *miombo* with woody vegetation have been used for construction, manufacturing, and the charcoal has formed a valuable fuel source. Many of the plants have been employed for traditional medicines and a wide variety of uses for these woodland savanna plants has been described. Colonization disrupted the traditional land-uses as witnessed elsewhere in Africa, and had the effect of concentrating population in urban areas and this in turn required a steady food supply from the savanna and the gradual establishment of more permanent settlements.

Over recent decades the pressure on *miombo* has increased and landscapes have been transformed, with much of the woodland converted into agricultural land or exotic tree plantations. Fires are also a major issue and controls have either been too weak or inadequately managed. The insecurity of land-tenure and the lack of effective policies have led to the *miombo* becoming a threatened part of the savanna.

Southern Africa. Land uses in South Africa and neighbouring Botswana and Namibia have reflected the seasonal rains and nutrient availability. Fires have also been a long continued feature in shaping the landscapes. The colonial period saw an end to the hunter-gathering economies of the indigenous San people although their domestication of cattle also has a long tradition. Small-scale farming was also widespread where water and sufficient fuelwood was available, as exemplified by the Bantu farmers. Colonization had many lasting effects, ranging from changing the pattern of land tenure and the creation of tribal reserves to the divisive policies of apartheid. Independence brought back a degree of communal control over land and redistribution of large landholdings to smallholders.

Nevertheless, the savanna landscapes have retained much of their original character but with the added inclusion of many large protected areas and state controls over land distribution. In contrast to the situation in the Americas, there is a substantial density of population in grassland or woodland savanna in southern Africa and this may be a reflection of the long history of land occupation. The two main forms of livelihood are livestock ranching along with subsistence cultivation and wildlife ecotourism. Pastoralism is believed to be prevalent over around 75 per cent of the savanna, mostly in extensive commercial farms but also on communal lands. The greatest numbers of domestic animals are found in the moist grasslands of South Africa, but substantial numbers of cattle and goats are kept in neighbouring Botswana and Namibia and there has been concern over the level of overgrazing. Much of southern Africa is too seasonally dry for cultivation but it is estimated that around 10 per cent of South Africa is farmed, mostly in the wetter and fertile parts of the savanna where maize is the principal crop. Fuelwood is the major source of energy in rural savannas.

Some of the main concerns at the present time are political issues connected with land management, including land tenure, overgrazing and bush encroachment, the state of conservation and wildlife, the increase in invasive plants, and fire management.

Landscape change in Australasia

Asia. While many of the savanna landscapes found throughout India and South-east Asia result from clearance and disturbance, there are a few remaining natural patches. These are maintained by burning, natural fires, or occur in seasonally flooded locations. Several types of woody savanna have been described from India eastwards and include some of the densest formations, typically of trees from the Dipterocarp family. Much of the present day savanna landscape is believed to be the relict of more extensive savanna vegetation that existed in the late Quaternary. The patches that remain have been the focus of increasing pressure from rapidly growing populations.

The land has traditionally been managed communally with strong ethnic links to the past and a sense of being trustees of the land to be passed on to successive generations. However, colonialism brought with it a different view of the use of wood resources, as timber for export, while favourable sites were developed for rice and other grains. Subsequent to independence in the differing nations of the region, there has been rapid legal and illegal deforestation of many of the savanna woodlands and increasing land conflicts, as illustrated in the Korat region of north-eastern Thailand. One example has been the clearance of wooded land in the upper catchments of streams leading to severe flooding. Demands for pulp and paper have also led to the planting of exotics such as eucalyptus species. The areas under any form of protection are scarce and vulnerable, with scattered tracts of savanna and increased population pressure from inward migration and resettlement schemes.

Amongst the concerns over land use that have been highlighted by commentators are political upheavals and uncertainty; deforestation, flooding, planting of exotics, construction of dams and infrastructure; mining and fires.

Australia. The continent possesses one of the widest ranges of savanna vegetation, from some of the driest to some of the wettest. However most of the savanna landscapes lie on undulating lower plains and there is less of the higher plateau topography typical of the Gondwana blocks in Africa and South America nor the volcanic landscapes of Central America and East Africa. Landscapes that can be described broadly as savanna cover a very large part of northern Australia and the variety of habitats has resulted in a valuable archive of land use research.

Over recent years there has been greater emphasis on the aboriginal way of life and how they managed the landscape. There is little doubt from some of the earliest records that fire was used as a tool to drive out game, for clearing vegetation cover to limit pests (mosquitoes and snakes, for example), and for cooking. This long period of anthropogenic fires coupled with those generated by lightning has resulted in many thousands of years of burning and plant adaptation. It has been suggested that some of the drier areas of savanna have been produced by this constant fire disturbance. Fire management by the aboriginal peoples reflected a good knowledge of the seasons and timing has been a key feature together with an intricate knowledge of how to set fires for maximum effect. The rich variety of savanna habitats provided a wealth of food resources. Well over a hundred species of edible plants were used together with a wide range of animals including fish, birds, and reptiles as well as mammals of all sizes. The knowledge of seasonal changes as well as an intimate understanding of the nuances of the environment meant that the indigenous peoples were highly mobile and well versed in utilizing all parts of their environment. The deep respect for land was reflected in their systems for tribal territories and land tenure. The impact of colonization was devastating to this way of life and the total population is believed to have dropped from around ¾ of a million to less than 100,000 before World War II. Groups were shepherded into reserves (around 3 per cent of the total area) on less favourable land. However, changes in attitude over the past few decades and laws such as those governing aboriginal land-rights have led to improved opportunities and more sympathetic integration.

The land use implanted by European colonization was essentially cattle-ranching developing into some of the largest farms in the world. Many are owned by large corporations which tend to be very large—some over 5,000 km². The introduction of better disease-resistant Brahman cattle during the 1960s improved the productivity. The other principal land-uses are for the military (the principal strategic reserve in the north of Australia) and a growing tourist industry that has strengthened the position of the protected areas. The present day situation is that the savannas are still very sparsely populated and support a very low percentage of the Australian population. Possibly a quarter of the inhabitants of the savanna landscapes are aboriginals.

Population growth and unequal distribution

One of the intriguing aspects of the world savanna landscapes is that they have been traversed relatively easily throughout history but did not act as magnets for permanent settlement. From the earliest times, they were utilized for foraging, hunting, and gathering, but only the most favourable sites with proximity to water or shelter were occupied continuously. There was relatively little cultivation and livestock-rearing of any type was mainly at a small scale. This pattern persisted until the period of European colonization when there was a demand for minerals and food resources. Urbanization became significant with the development of industries and transport infrastructure in the 19th century. So the density of people over what may strictly be defined as savanna (as opposed to the savanna landscape as a whole) has been very low in the past. In today's world it can be argued that the savannas represent some of the largest tracts of relatively undeveloped and unoccupied yet potentially valuable land and are therefore under growing threat of exploitation. The concerns raised by this will be explored further in the last chapter but here the emphasis will be upon the way in which the population has grown.

It is difficult to arrive at accurate figures for the numbers of people living in savannas because the census data follow political and not natural boundaries. The global estimates mentioned in the first chapter embrace the complete mosaic of constituent savanna ecosystems and include people who depend upon savannas for their livelihood but who do not necessarily live permanently in them. One of the ways to illustrate such trends is to consider typical savanna parts of the countries within the savanna biome. The United Nations figures for world population and urbanization prospects reveal a striking growth in the proportion of people living in urban areas. Despite the rural nature of much of the savanna today, population increase is predicted to be mostly in the urban settlements of the less developed regions of the world. As a broad generalization, Latin American and Caribbean countries are already highly urbanized (over 75 per cent), with Africa and Asia lower at around 40–50 per cent.

The increases in urbanization reflect the global patterns of population growth. For example, the population of Brazil has risen from just over fifty million in 1950 to over 180 million only fifty years later. It is predicted to rise to around 220 million by 2050. India, which contained very large expanses of savanna in prehistoric times, has seen a population rise from around 370 million in 1950 to 1,130 million fifty years later, with UN predictions of 1,614 million by 2050. There is little left of the original landscape except for a few isolated marginal areas such as the Terai in the north close to Nepal. What appears today as savanna-like vegetation is an anthropogenic landscape. Some of the African savanna-dominated countries show some of the most significant increases in the numbers of people. For instance, in West Africa, Burkina Faso has seen a population increase from four million in 1950 to over fourteen million (2005), and Niger from 2.5 million to thirteen million in the same period. Both of the countries are predominantly in the seasonally very dry savanna regions with severe limitations on cultivation and the pastoral activities. In Eastern Africa, Tanzania makes an equally striking case study with a rise from 7.5 million in 1950 to over forty million today, and a UN predicted rise to 110 million by 2050. Botswana, with a very sparse population in a savanna region not considered favourable for agricultural development, jumped from under half a million in 1950 to nearly two million some fifty years later, with a predicted rise to over three million by 2050. While these examples parallel the growth of national populations and urbanization generally, they are significant because they are located in savanna landscapes which until recently have not been shown to support such levels of rural density. The situation is even more tenuous in the drier savannas with less reliable rainfall and episodic severe drought.

The rise of savanna cities

While rural areas in savanna have tended to be lightly populated, some surprising and remarkable concentrations of people are concentrated in their urban areas. Most of the settlements have been located for political, strategic, or economic reasons, or lie close to water sources.

Brasília is perhaps the prime example of a savanna city, although its location depended upon the damming of rivers to form the large lake of Paranoá. The site selected for the new capital was in the centre of the country and was expected to act as a focus for economic growth in the undeveloped west. Although the city was conceived as far back as 1827 and planned in the mid-1950s, the official foundation was in 1960. By 2014 the population had rocketed to 2.8 million with a number of planned but more spontaneous satellite cities growing outside the pilot plan, such as Sobradinho (*c.*130,000), Taguatinga (*c.*24,000), or Ceilandia (now estimated to hold around 350,000 people) (see Figure 32).

The new capital took over from Rio de Janeiro (capital from 1763 to 1960) and in the early years there was considerable reluctance to move from the vibrant life on the coast to the quiet life in the virtually untouched *cerrado*. The Federal District was carved out of the state of Goias, which illustrated an earlier manifestation of a similar process of expansion. The original capital of Goias was Goias Velho founded, like so many interior towns, by the earlier pioneers (*bandeirantes* or flag carriers), but in 1933 the state capital was moved to the planned city of Goiania in a more strategic position and this was boosted with the rise of Brasília only 200 km away. The city prides itself on being green and is surrounded by the *cerrado* and gallery forests with some of the typical buriti palm groves (*Mauritia flexuosa*) along damper depressions. However, the spectacular advance of cultivation and livestock, particularly cattle farming, has already converted

**32. Brasília—a savanna city. The Government Buildings and
Parliament (in the middle distance) were implanted in the sparsely
populated savannas of central Brazil. The site was chosen where
a shallow valley could be dammed to create a water reservoir
(Paranoá).**

much of the original landscape. By 2014 the population had risen
to 1.5 million out of a state containing 6.5 million people. Other
urban areas implanted into the savanna landscape in South
America include examples such as San Fernando in Apure State in
the Venezuelan *llanos*. The city was founded by missionaries in
1788 on the banks of the Apure River, in a very flat part of the
grassland. The population has now risen to around 200,000,
dominated by cattle-rearing and agriculture. The relative distance
of some savanna cities from the highly urbanized capitals may
have limited development and this seems typical of the furthest
savanna regions, reached by the movement of the agricultural
frontier but not spurred by political or industrial stimuli.

The African savanna cities may be thought to have originated
much earlier, along with the lengthy history of human occupation.
However, early settlements were relatively small-scale, even the
city states of West Africa or the estimated 10,000–20,000 people

in Greater Zimbabwe, the largest pre-colonial city south of the Sahara. The large cities such as Nairobi or Harare can be traced back to fairly recent political and economic changes and capital cities have their own national momentum. Nairobi is named from the Masai word for 'cool water', and points to its location by the Nairobi River. It is situated near the eastern edge of the Rift Valley and is surrounded by open plains with dry forest on higher land to the west. The presence of a National Park occupying around 120 km^2 only a few kilometres from the city centre conveys a striking atmosphere of savanna. Like Brasília, it has a plateau elevation, in this case nearly 1,800 m above sea level. In 1899 it developed as a depot on the rail link between Mombasa and Uganda. Yet today it is estimated to have a population approaching four million and this has triggered enormous pressures on communications, commerce, and housing. These developments have inevitably spilled out over the surrounding countryside and shaped the landscape utterly.

Harare at around 1,500 m above sea level has an analogous history. The city is situated in the north-east of the country on the Highveld plateau and lies within the characteristic *miombo* savanna woodlands. The indigenous Shona people had a village at the site which gave the capital its name, but the city grew up largely as a transport hub from its foundation in 1890. By 2014 the population had risen to over 1.6 million with 2.8 million living within the metropolitan area. The political landscape has changed as has the savanna since independence, and the land use of the surrounding *miombo* is in a state of flux with transformations in land tenure and land ownership and consequently in the nature of agriculture and use of the woodlands. Where there is sufficient water, the mosaic of savanna landscapes has been shown to be capable of producing high yields of staples such as maize and export crops such as tobacco and cotton. There have been plans to develop a new capital some 20 km to the north-west, partly to relieve pressure on Harare but also a political move, so it will be particularly interesting to see what directions urbanization will take in the future. Further south, Johannesburg and Gaborone reflect locations within savanna but with quite different origins.

In West Africa the largest concentrations of urban population are found in the more moist areas of the Guinea Zone and there are fewer inhabitants of dry savannas such as the Sahel and parts of the Sudan zones. An example of a savanna city is Ouagadougou in Burkina Faso, a country slightly larger than New Zealand, where 80 per cent of the seventeen million people live in rural areas. Nevertheless, the capital contains over 1.5 million people and lies on the central plateau in the northern Sudan zone savanna. The country is mostly undulating and level and the key to the location of the capital is again water supply—lying in the basin of the Niger River which drains around a third of the national area.

The situation in South Asia, South-east Asia, and Australia reflects similar processes but different outcomes. India has many large urban centres in what is likely to have been savanna but population growth and landscape transformation make the original links with savanna difficult to trace. In northern Australia the mainly rural life has not given rise to major urban centres. However, mining has triggered a rapid urbanization in the Queensland interior (such as Mt Isa) and a few smaller urban centres exist at transport nodes such as Katherine. Like so much of Australia the principal urban centres have been the political state capitals, and Darwin represents both state capital and coastal savanna city. It currently holds around 150,000 people but for a long time was isolated from the rest of the country. World War II stimulated growth because of its strategic position on the Timor Sea, and today the population is still boosted by the nearby military town of Palmerston. However, the economy is dominated by mining (gold, zinc, bauxite, oil, and gas), together with tourism, which makes good use of the spectacular environments of the Kakadu National Park and Lichfield with in a few hours' drive from Darwin.

Savannas

The growth of cities in the savanna and the greatly increased pace of urbanization have placed enormous pressures on a land that witnessed relatively little settlement before the last few decades, and arguably where the developments are largely unconnected with the landscape.

Chapter 7
Savanna futures

Constant change is the driving force in the history of savannas, and some conception of the future can be gauged from evidence of past changes and from trends experienced today. Geological history reveals major global and regional climatic changes, the most recent of which include glacial advances and retreats that influenced savannas even though their greatest impact was in higher latitudes. Since the last glacial retreat around 20 Kya and within the historical record, there have been oscillations of temperature and rainfall that have come to light from sediment cores, evidence of past vegetation in pollen or diatom profiles, and in rock paintings and other archaeological evidence. Physical and biological processes that are currently affecting savanna regions include persistent changes in climate, especially rainfall patterns and temperature increases; reduced discharge of rivers and lowering of water tables; shifts in the ranges of plants, their phenology, and growth; losses in biodiversity; declining crop yields and desertification.

The pace of these changes is occurring faster than that considered as the result of natural variation. Whereas some climatic and environmental changes can be anticipated from knowledge of these trends, many human interventions are relatively new within savannas and introduce pressures whose long-term effects are unknown. Our stewardship of these resources depends on clear information and assessment, and a combined approach integrating the interests of all those who live in or utilise savanna resources.

The potential futures for savanna landscapes may be depicted as two interconnected sets of processes: variations caused by natural events, and changes resulting from human activities.

Climatic change and its implications

Climates have varied over time and differ in each regional location, but there is consistent evidence that present day global variations are having repercussions throughout all savannas. The report from the Intergovernmental Panel on Climatic Change (IPCC) in 2014 tried to distinguish human impacts from those caused by natural changes. It comes to the unequivocal opinion that anthropogenic pressures are accelerating climatic change. The principal directions are increases in global temperatures and changes in atmospheric composition, which have knock-on effects for the environment and on human societies. Long-term effects that particularly influence savannas include accentuating seasonality with more extreme temperatures and rainfall patterns, together with changes in atmospheric CO_2 concentration. More immediate changes are likely to result in increased storm intensity and variations in global influences such as the ENSO events or the ITCZ. Natural climatic changes have usually been over a period long enough for vegetation to adapt or diversify, and for animals to have a chance to migrate, whereas human actions are more immediate and can have radical impacts on climate, land use, and wildlife.

The IPCC estimate that it is likely global surface temperatures will increase by 1.5°C (with some projections suggesting 2°C), by the end of the 21st century relative to the situation from 1850 to 1900. While these temperatures lead to moderate impacts, a rise of 3°C would result in marked losses in biodiversity and the benefits offered by what are grouped as ecosystem services. The impact of temperature rises beyond such levels falls within the realms of informed guesswork but almost certainly with a much worse scenario. Global warming is largely caused by the emission of greenhouse gases such as carbon dioxide, nitrous oxide, and methane. Variations in the global water cycle are predicted to increase disparity between warm and wet regions, and between wet and dry seasons, with distinctions between different regions. Much of the increased energy is likely to be dissipated into the oceans, leading to thermal expansion, which, together with the melting of ice sheets, is forecast to affect low-lying savanna coastlines through rising sea levels.

The probable mean sea level rise from 1900 and 2010 was 1.7 mm y^{-1} according to the IPCC, but rose to 3.2 mm y^{-1} between 1993 and 2010. Future predictions to 2100 suggest a rise of between 0.44 m and 0.74 m depending on the level of emissions. By 2200 this forecast rises to well over 1 m. These estimates exclude storm surges and what are termed singular events (such as the total deglaciation of Greenland). These figures may not seem at first to be very urgent, but there are many savanna areas that are close to the coast and would experience increased flooding and salt water incursions, including many vulnerable settlements. For example, the height of Belize City in Central America is only around 20 cm above mean sea level, and that figure itself has been whimsically referred to as being built on 'mahogany chips and broken gin bottles' from its colourful past. Much larger cities have also been shown to be vulnerable, such as São Paolo in Brazil, located at the eastern edge of the *cerrado*, or Darwin on the margins of the northern Australian savannas. Coastal savannas are widespread around the Caribbean and Australia and are to some extent protected by fringing mangrove that, in contrast to savanna vegetation, seems able to cope with gradual changes in sea level.

Increased concentration of CO_2 in the atmosphere affects plant growth by influencing photosynthetic rates and this has sparked a lively debate as to its role in savanna vegetation. Concentrations of CO_2 in the atmosphere have varied considerably over time: grasslands expanded during low CO_2 concentrations, as during glacial periods, and forests amplified during interglacials. Experimental and simulation studies predict increases in the number of woody species and related biomass as a result of greater atmospheric CO_2, assuming there is sufficient moisture. The advance of woody species into grassland can be correlated with the history of industrialization and gas emissions, as shown in studies from South Africa. At the same time there are predictions of savanna expansion at the forest fringes with the onset of drier climates, especially where coupled with land use changes (as forecast for eastern Amazonia). Vegetation transformations in more arid savannas appear to be less clearly linked to CO_2 and may be driven by over-grazing. Some dry savannas seem to have remained stable over decades such as the Australian woody savannas and open grasslands. It is likely that CO_2 levels will increase beyond what has been described as the 'evolutionary experience' of savanna plants, and therefore that we could witness a decrease in C4 plants and a loss of grass-dominated landscapes. Nevertheless modest increases in CO_2 could be argued to be beneficial overall, since vegetation stores more carbon resulting in greater biomass. The upper limits of this photosynthetic balance are far from clear.

Climatic change also affects the impact of fires. With seasonally increased temperatures, greater evaporation, and diminishing soil moisture, fires are able to override climatic influences in places where there is sufficient combustible material built up during wetter periods. The humid savannas are generally more productive and thus generate a greater quantity of organic matter, whereas more arid savannas have fewer fires because they lack as much combustible material. Savanna trees are generally distinct from evergreen forest, in function as well as species composition, and they adapt to burning except in extreme conditions. High fire incidence favours the development of underground storage tubers and other protective measures as well as affecting species composition. Savanna soils act as important C-sinks and biomass changes above ground are likely to be reflected in as yet inadequately quantified subsurface variations. Following disturbances such as fires, an increased CO_2 concentration would favour tree growth over grassland. The colonization of open areas by trees is also helped where woody plants can expand from wetter sites such as galleries or damp depressions. Where fire frequencies are restricted, there has been widespread expansion of woody vegetation responding to climate over recent decades in many but not all savanna regions. Despite increasing information on these trends, it is extremely difficult to predict where the changes may occur due to the interaction of a multitude of factors, and this makes management of savannas a complex and multiple-scaled endeavour.

Human activities and their implications

Many of the savanna landscapes have already been greatly disturbed or completely transformed as depicted in Chapter 6. Trends portray sweeping shifts in land use that affect not only the plants, animals, and human settlements dependent upon savanna resources, but also impact on the global economy. Reports from international agencies such as the UN (Millennium Ecosystem Assessment), IPCC (Climatic Change), or the WWF (Ecosystem Assessments) amongst numerous other groups, concur on the need for an integrated approach to dealing with the concerns encompassing both biological and social issues, but warn of factors that are threatening solutions.

The transformation of natural savanna landscapes has resulted from a relatively small number of major activities. Possibly the most widespread activity is pastoralism, which has taken place for thousands of years in parts of the African savanna but which has reached more commercial proportions in modern farming practices. Some of the New World and Australian livestock farms occupy enormous areas, with the concomitant dangers of monoculture such as lack of biological diversity and buffering against pests and diseases. Similarly increasing cultivation for cash crops has limited the resource base of savanna regions across the globe and resulted in monotonous, biologically limited landscapes (see Figure 30). Despite this increasing pressure on land, there are still many areas with resident indigenous people practicing smallholder farming, some of whom live in protected areas and National Parks. More exotic land uses include mining, which takes on a different character depending on the product and location, but is always exploitative at the expense of the environment, along with the urban development and expansion of infrastructure. Over recent years there has also been growing recognition of the wider range of ecosystem services that the savannas can offer, pointing to the value of environmental, social, cultural, and aesthetic benefits beyond economic resources. There have been many attempts to put a value on nature, but not much consensus as to the best way forward. Whilst it can be argued that many of the activities are necessary for the economic well-being of the countries, it is equally obvious that a balance needs to be struck between transformation, addressing social and economic aspirations, and conserving the biological resources.

Balancing the processes

The main concerns that could impact upon the savannas have been identified by the IPCC and other international and local agencies. They can be reduced to a number of key issues.

Coastal savanna areas. These are identified as being at risk of flooding and salt-water incursions. Yet several developments help to alleviate these problems. The greatly increased accuracy of weather forecasting from satellite data has enabled early warning to be given to vulnerable communities, such as the hurricane warnings in Middle America or cyclones in northern Australia. Similarly the level of information and media communication (through television, radio, and mobile phones) has reached out to almost all communities in what were previously isolated and remote areas.

Food security and food production systems. Increased desiccation threatens all plant growth and agriculture has always been a challenge for vulnerable areas in savanna regions. The wetter savannas have sufficient water but lack nutrients, whilst the dry areas often have more potentially fertile soils but lack water. Land degradation through processes such as leaching and soil erosion frequently results from the depletion of land cover. Land surveys and monitoring from satellites with multispectral sensors at practical scales using GIS now provide more accurate and up-to-date information. Many of the problems appear to have a measure of technical solution (fertilization, organic matter management, restoration, soil moisture control). However, these are usually costly and beyond the reach of smallholder farmers, who rarely have access to credit, and most assistance seems to be directed towards the larger producers. The development of new and modified crops, better storage and marketing, and multiple land use schemes could potentially help all scales of farming. Conflicts over land tenure are an endemic feature of many tropical areas and in some countries land registration surveys have alleviated some of the disputes but there are many instances where land is claimed by several competing owners. The involvement of international agencies such as the FAO or CGIAR (representing an influential group of agencies), and the increasingly active role of NGOs, has significantly helped to relieve the problems of food security with better information and communication of techniques.

Loss of biodiversity. This is a widespread issue throughout savannas and is scrutinized in a recent report to the EU concerning the worsening situation in Africa (see the citations in websites given in the 'References and further reading' section). Whereas most of the countries in the sub-Saharan Africa are signatories to the various international environmental conventions and IUCN outlines, there are widely divergent levels of implementation. The report identifies trends which can be mirrored in most of the world's savanna regions—a steady erosion of plant and animal wildlife along with increased pressure on land and natural resources. The report points to the value of protected areas (PAs), which represent the most intact assemblages of African wildlife, but it also recognizes that many indigenous groups consider such reserves to be imported concepts that should not interfere with traditional ways of life. It is clear that for PAs to flourish there have to be tangible benefits for local people. Environmental- and eco-tourism have offered some opportunities, where people are compensated for damage in return for protecting wildlife or where there is significant tourist revenue. Other initiatives that potentially offer some return for local people include extending the UN REDD (Reducing Emissions from Deforestation and Forest Degradation) proposals to savanna woodlands and more open savannas, especially if the value of soil organic storage becomes better appreciated. The most promising approaches seem to be strengthening the National Parks and other PAs through long-term funding, especially from wealthier external stakeholders, with better reserve design and connecting corridors. The Transfrontier Conservation Areas, first initiated as the South African Peace Park in the 1990s, is currently being expanded with a possible eighteen potential schemes. Public and Private Partnerships also offer means of engaging with local people and incorporating monitoring and research.

Conservation issues. The arguments surrounding conservation and development stand out amongst global concerns. They are well illustrated in the Brazilian *cerrado* where well over 50 per cent of the vegetation has already been transformed into planted pastures and agricultural cash crops. The rate of deforestation is greater than that in the much more widely publicised Amazonian forest, and less than 3 per cent of the area is under any legal protection. This has entailed a high environmental cost with fragmentation of remnant vegetation, soil erosion, increased leaching and soil compaction, water pollution, greater frequencies of fires, and imbalances to the C-cycle. Many habitats are sensitive to change affecting many of the essential pollinators such as bees.

The loss of biodiversity is exemplified by the elimination of native varieties of cassava (*Manihot* species), which is one of the world's major food crops. On a more positive note, a number of alleviating forces have been at work and are growing in influence. While it would be unrealistic to imagine that the conversion of land into farming will cease, there is a better awareness of the importance of retaining biodiversity, better understanding of agricultural practices, and the adoption of different techniques such as minimum tillage and agroforestry. There is also growing awareness of the value of indigenous knowledge and of the contribution from different ethnic groups and opinions. Introduction of some of the invasive African forage grasses is also being closely monitored including *Melinis minutiflora*, which has a particularly disruptive impact on native herbaceous communities. There has been a growing influence from a network of NGOs, and some state governments, notably Goias, have been active in promoting more sustainable measures. The protected areas are in danger of being surrounded by converted land and so there are plans to develop wildlife corridors (such as between the *cerrado* and the Pantanal). The greatest advance, however, has possibly been the worldwide recognition that savanna biodiversity is important ecologically, culturally, and economically, and that there is widespread publicity to conserve as much of the natural resources as possible.

Urban expansion. The rapid increases in the proportions of people living in towns and cities are transforming many of the savanna landscapes. Many urban immigrants live in what are euphemistically termed 'informal settlements' (i.e. squatter districts). The greatly increased population and growth rates in sub-Saharan Africa are largely directed to the urban areas. The movement from rural to urban areas is having a significant impact on savannas and the trend is unlikely to diminish. Once again international agencies have been active in promoting schemes for integration, despite this being perceived as a 'top-down' approach. In the end it becomes a national issue requiring political solutions.

Human health and security. Problems associated with these issues have been accentuated with increased temperatures, less predictable rainfall, and greater numbers of storm events. An upsurge in abnormal climatic events and their environmental consequences has alerted international communities to the need for practical measures and rapid response mechanisms. The occurrence and intensity of such eventualities varies widely over the savanna, but for example Central America is picked out as being 'one of the most responsive tropical regions to climatic change'. Coastal lowland savannas are clearly at greater risk. A number of proposals have been made to counter the causes of threats with a number of international agencies involved such as the UN and FAO. The most immediate problems have been identified as population growth and poverty; poor governance and/or weak legislation and enforcement; inadequate land tenure and resource rights; conflicts and violence; water shortages; disease and political indifference.

Ultimate futures

A balanced approach would ideally embrace the whole of the savanna landscape rather than individual components. Over the broad canvas of contemporary ideas, there are many practical suggestions for limiting population growth (partly by raising living standards) and focusing on the key elements essential for conservation. There is a strong case for a Key Landscape Conservation approach, together with the African model of Transfrontier Conservation Areas and designation of individual sites of outstanding value. It is evident that any form of sustained conservation, however well integrated and inclusive, is going to require stronger administrative controls and legal frameworks. At the same time it will be necessary to tackle corruption and criminal activities, whilst dealing with issues such as bushmeat and fuelwood which are part and parcel of the livelihoods and traditions of many people. Since the landscape and the political boundaries rarely coincide, it is often necessary to examine individual aspects separately and then to seek integration. It is also usually necessary to compromise between opposing interests and to find the 'Middle Ground', as eloquently advocated by Reid for East Africa.

While many of the current concerns are to some extent being addressed, and the worst impacts of natural and human change better understood, most measures are mitigations not solutions. Forecasting the fate of savannas is a risky proposition, yet the trends outlined are likely to persist and probably magnify. It remains difficult to be optimistic about future alternatives for the savannas when political instability, population growth, and demands on the land continue to outweigh resources. The concept of a savanna landscape is evolving from one that is natural to one that is manipulated. In all likelihood it will become a totally anthropogenic landscape and therefore one where we cannot escape responsibility for its husbandry. Savannas are landscapes to rejoice in and wonder at but they also need to provide a livelihood as well as a world resource.

Savannas

References and further reading

General texts

F. Bourliere (ed.), *Tropical savannas: ecosystems of the world*, vol. 13 (Amsterdam: Elsevier, 1983). A comprehensive overview of the savanna biome.

M.J. Hill and N.P. Hanan (eds), *Ecosystem function in savannas: measurement and modelling at landscape to global scale* (Boca Raton, FL: C.R.C. Press, 2011). Chapters dealing with a number of general concepts as well as processes and technical issues, including gas and water fluxes, remote sensing, and modelling.

B.J. Huntley and B.H. Walker (eds), *Ecology of tropical savannas* (New York: Springer-Verlag, 1982). General overview of the biological aspects of savannas.

P.L. Osborne, *Tropical ecosystems and ecological concepts* (Cambridge: Cambridge University Press, 2012, 2nd edn). Useful introduction to the place of savannas within the tropical biomes.

J. Mistry, *World savannas* (Edinburgh: Pearson, 2000). A readable introductory volume covering both the ecological and the human use of savannas.

Specific references and texts

Chapter 1: Savanna landscapes

C.E.R. Lehmann, S.A. Archibald, W.A. Hoffman, and W.J. Bond, Deciphering the distribution of the savanna biome. *New phytologist* 191, 197–209 (2011).

C.E.R. Lehmann, J. Ratnam, and L.B. Hutley, Which of these continents is not like the other? Comparisons of tropical savanna systems: key questions and challenges. *New phytologist* 181, 508–11 (2009).

J. Ratnam, W.J. Bond, R.J. Fensham, W.A. Hoffman, S. Archibald, C.E.R. Lehmann, M.T. Anderson, S.I. Higgins, and M. Sankaran, When is a 'forest' a savanna, and why does it matter? *Global ecology and biogeography* 20, 653–60 (2011).

Chapter 2: Shaping the savannas

A.N. Andersen, G.D. Cook, and R.J. Williams (eds), *Fire in tropical savannas: the Kapalga experiment* (London: Springer, 2003).

D.E. Bignell, Y. Roisin, and N. Lo (eds), *Biology of termites: a modern synthesis* (Dordrecht: Springer, 2010).

M.A. Cochrane, *Tropical fire ecology: climate changes, land use and ecosystem dynamics* (Berlin: Springer/Praxis, 2009).

P.A. Furley, R.M. Rees, C.M. Ryan, and G. Saiz (2008), Savanna burning and the assessment of long-term fire experiments with particular reference to Zimbabwe. *Progress in physical geography* 32, 611–34.

J. Mistry, The ecosystem dynamics of tropical savannas, in A. Millington, M. Blumer, and U. Schikoff (eds), *The Sage handbook of biogeography* (London: Sage, 2011), pp. 281–302.

F.M.S. Moreira, E. Moreira, H. Huising, and D.E. Bignell, *A handbook of tropical soil biology* (London: Earthscan, 2008).

O.T. Solbrig, E. Medina, and J.F. Silva, *Biodiversity and savannah ecosystem processes* (New York: Springer, 1996).

B.H. Walker (ed.), *Determinants of tropical savannas* (Oxford: IRL Press, 1987).

Chapter 3: Savanna vegetation

B. Campbell (ed.), *The miombo in transition: woodlands and welfare in Africa* (Bogor, Indonesia: Center for International Forestry Research, CIFOR, 1996).

E.N. Chidumayo, *Miombo ecology and management: an introduction* (Stockholm: Stockholm Environment Institute, 1997).

R.M. Cowling, D.M. Richardson, and P.M. Pierce (eds), *Vegetation of southern Africa* (Cambridge: Cambridge University Press, 1997).

P.A. Furley, J. Proctor, and J.A. Ratter (eds), *Nature and dynamics of forest-savanna boundaries* (London: Chapman & Hall, 1992).

Savannas

J. Grace, J. San Jose, P. Meir, H. Miranda, and R. Montes, Productivity and carbon fluxes of tropical savannas. *Journal of biogeography*, 387–400 (2006).

C.E.R. Lehmann and others (21), Savanna vegetation-fire-climate relationships differ among continents. *Science*, 343, 548–52 (2014).

R.T. Pennington, J.A. Ratter, and G.P. Lewis (eds), *Neotropical savannas and seasonally dry forests: plant diversity, biogeography and conservation* (Boca Raton, FL: CRC Press, 2006).

P.S. Oliveira and R.J. Marquis (eds), *The cerrados of Brazil: ecology and natural history of a neotropical savanna* (New York: Columbia University Press, 2002).

G. Sarmiento, *The ecology of tropical savannas* (trans. O. Solbrig; Cambridge, MA: Harvard University Press, 1984).

Chapter 4: Wildlife and microbes

D.E. Bignell, Y. Roisin, and N. Lo (eds), *Biology of termites: a modern synthesis* (Dordrect: Springer, 2010).

R.D. Estes, *The gnu's world* (Berkeley: University of California Press, 2014).

R.T. Corlett, *The ecology of tropical East Asia* (Oxford: Oxford University Press, 2014, 2nd edn).

P.S. Oliveira and R.J. Marquis (eds), *The cerrados of Brazil: ecology and natural history of a neotropical savanna* (New York: Columbia University Press, 2002).

R.J. Scholes and B. H. Walker, *An African savannah—Nylsvley* (Cambridge: Cambridge University Press, 1997).

B. Shorrocks and W. Bates, *The biology of African savanna* (Oxford: Oxford University Press, 2015, 2nd edn).

A.R.R. Sinclair, *Serengeti story* (Oxford: Oxford University Press, 2012).

Chapter 5: Savannas and human evolution

R. Boyd and R.B. Silk, *How humans evolved* (New York: W.W. Norton, 2015, 7th edn).

S. Cane, *First footprints: the epic story of the first Australians* (London: Allen & Unwin, 2013).

W.M. Denevan, *Cultivated landscapes of native Amazonia and the Andes* (Oxford: Oxford University Press, 2001).

C. Higham, *Early mainland south East Asia: from first humans to Ankor* (Bangkok: River Books, 2014).

M. Jobling, E. Hollox, M. Hurles, T. Kivisild, and E. Tyler-Smith, *Human evolutionary genetics* (London: Garland Science/Taylor & Francis, 2014, 2nd edn).

T. M. Whitmore and B.L. Turner, *Cultivated landscapes of Middle America on the eve of conquest* (Oxford: Oxford University Press, 2001).

K.J. Willis and J.C. McElwain, *The evolution of plants* (Oxford: Oxford University Press, 2014).

Chapter 6: Changing patterns in the landscape

C.J.R. Alho and E de S. Martins (eds), *Bit by bit the cerrado loses space* (Brasília: WWF. PRO-CER, 1995).

T.J. Bassett and D. Crummey (eds), *African savannas: global narratives and local knowledge of environmental change* (Oxford: James Currey, 2003).

C.A. Klink and R. B. Machado (2005), Conservation of the Brazilian cerrado. *Conservation biology*, 19, 707–13.

P.K.R. Nair, *An introduction to agroforestry* (Dordrect: Kluwer, 1993), particularly section IV: Soil productivity and protections.

P. Werner, (ed.) *Savanna ecology and management: Australian perspectives and intercontinental comparison* (Oxford: Blackwell, 1991).

M.D. Young and O.T. Solbrig (eds), *The world's savannas: economic driving forces, ecological constraints and policy options for sustainable land use. Man and the Biosphere*, vol. 12 (Paris: UNESCO/Parthenon Press, 1990).

Chapter 7: Savanna futures

J. Mistry and A. Berardi (eds), *Savannas and dry forests: linking people with nature* (London: Ashgate, 2006).

R.S. Reid, *Savannas of our birth: people. Wildlife and change in East Africa* (Berkeley: University of California Press, 2012).

Websites and data bases

<http://capacity4dev.ec.europa.eu/b4life/minisite/biodiversity-life-b4life/african-wildlife-conservation-strategy> *Larger than elephants: inputs for the design of an EU strategic approach to wildlife conservation in Africa.* Several sections (2015) covering all parts of Africa, reviewing and assessing wildlife and conservation issues.

<www.maweb.org/index.aspx> Millennium Ecosystem Assessment.

<www.ipcc-wg2.gov/AR5/report/> The detailed Intergovernmental Panel on Climate Change report 2014.

<www.openknowledge.worldbank.org> World Bank group reports (IBRD, IDA, IFC, MIGA, ICSID) containing social as well as economic reviews.

<www.panda.org/about_our_earth/all_publications/living_planet_report> WWF reports covering many aspects relevant to savannas and wildlife.

<www.fao.org/soils-portal/soil-survey/soil-classification/en/> Describes the FAO world soil classification system.

<www.savanna.org.au> A helpful and illustrated account of the environment, vegetation, and wildlife of Australian savannas.

<www.nrcs.usda.gov/wps/portal/nrcs/main/soils/survey/class/taxonomy/> Describes the US Soil Taxonomy classification (2014) of the Soil Survey Staff, giving the keys to soil identification.

<www.cgiar.org/resources/cgiar-library/> Information and publications from the CGIAR group (Consultative Group for International Agricultural research), covering a wide range of scientific, political, and philanthropic organizations with research programmes on water land and ecosystems, climate change, agriculture, and food security.

<www.ccafs.cgiar.org> Information and publications of the Research Programme on Climate Change, Agriculture and Food Security. Led by the International Center for Tropical Agriculture (CIAT) and part of the collaboration between fifteen CGIAR research centres.

Index

SOCIAL MEDIA
Very Short Introduction

Join our community
www.oup.com/vsi

- Join us online at the official Very Short Introductions **Facebook** page.
- Access the thoughts and musings of our authors with our online **blog**.
- Sign up for our monthly **e-newsletter** to receive information on all new titles publishing that month.
- Browse the full range of Very Short Introductions online.
- Read **extracts** from the Introductions for free.
- If you are a teacher or lecturer you can order inspection copies quickly and simply via our website.